Die Grundlehren der mathematischen Wissenschaften

in Einzeldarstellungen
mit besonderer Berücksichtigung
der Anwendungsgebiete

Band 182

J. L. Lions · E. Magenes

Non-Homogeneous Boundary Value Problems and Applications

Translated from the French by
P. Kenneth

Volume II

Springer-Verlag Berlin Heidelberg New York 1972

J. L. Lions
University of Paris

E. Magenes
University of Pavia

Title of the French Original Edition:
Problèmes aux limites non homogènes et applications (tome II)
Publisher: S. A. Dunod, Paris 1968

Translator:
P. Kenneth
Paris

Geschäftsführende Herausgeber:
B. Eckmann
Eidgenössische Technische Hochschule Zürich
B. L. van der Waerden
Mathematisches Institut der Universität Zürich

AMS Subject Classifications (1970)
Primary 35J20, 35J25, 35J30, 35J35, 35J40, 35K20, 35K35, 35L20,
Secondary 46E35

ISBN-13: 978-3-642-65219-6 e-ISBN-13: 978-3-642-65217-2
DOI: 10.1007/978-3-642-65217-2

Softcover reprint of the hardcover 1st edition 1972

Library of Congress Catalog Card Number 71-151407

Introduction

1. In this second volume, we continue at first the study of non-homogeneous boundary value problems for particular classes of evolution equations.

In Chapter 4[1], we study parabolic operators by the method of Agranovitch-Vishik [1]; this is step (i) (Introduction to Volume I, Section 4), i.e. the study of *regularity*. The next steps: (ii) *transposition*, (iii) *interpolation*, are similar in principle to those of Chapter 2, but involve rather considerable additional technical difficulties.

In Chapter 5, we study hyperbolic operators or operators well-defined in thesense of Petrowski or Schroedinger. Our regularity results (step (i)) seem to be new. Steps (ii) and (iii) are analogous to those of the parabolic case, except for certain technical differences.

In Chapter 6, the results of Chapters 4 and 5 are applied to the study of *optimal control problems* for systems governed by evolution equations, when *the control appears in the boundary conditions* (so that non-homogeneous boundary value problems are the basic tool of this theory). Another type of application, to the characterization of "all" well-posed problems for the operators in question, is given in the Appendix. Still other applications, for example to numerical analysis, will be given in Volume 3.

2. For the same systems $\{P, Q_j\}$, Volume 3 will proceed in an analogous way, but starting from regularity results in *real analytic* or *Gevrey* classes. In this way, we shall be able to reach the case in which the data f, g_j are Gevrey functionals or analytic functionals.

3. The organization of the chapters is the same as for Volume 1. In particular, the open problems are assembled in the last section of each chapter.

J. L. Lions E. Magenes

[1] The numbering of the chapters follows that of Volume I.

Contents

Chapter 4

Parabolic Evolution Operators. Hilbert Theory

Chapter 5

Hyperbolic Evolution Operators, of Petrowski and of Schroedinger. Hilbert Theory

Chapter 6

Applications to Optimal Control Problems

Appendix

Boundary Value Problems and Operator Extensions

Contents of Volume I

(Published 1972)

Contents of Volume III

(In preparation)

Chapter 4

Parabolic Evolution Operators. Hilbert Theory

In this chapter we develop the theory of boundary value problems for partial differential equations of the parabolic type; this theory is in a way the analogue of the non-variational elliptic theory developed in Sections 1–8, of Chapter 2[1].

We shall not go back over the concrete applications of the results of Chapter 3 to parabolic equations, for which we have already given examples (see Section 5.2, Chapter 3). This chapter (except for Section 17.2) does not require the knowledge of Chapter 3.

1. Notation and Hypotheses. First Regularity Theorem

1.1 Notation

Let Ω be an open set in $\mathbf{R}^n$; we assume Ω to be bounded and to have an $(n-1)$ dimensional, infinitely differentiable boundary Γ such that Ω is locally on only one side of Γ (see Chapter 1, Section 7).

Remark 1.1. The following results can be extended to the case where Ω is not bounded but where Γ has a "suitable" behavior at infinity. We have not tried to extend what is to follow (as was not done in Chapters 2 and 3 either) to the case $\Gamma = \bigcup_j \Gamma_j$, where the dimension n_j of Γ_j may be *strictly* less than $(n-1)$; see Problem 17.7. □

Let $T < \infty$ and

$$Q = \Omega \times]0, T[,$$

$$\Sigma = \Gamma \times]0, T[, \qquad \textit{lateral boundary of } Q$$

$$\Omega_0 = \Omega \times \{t = 0\}$$

$$\Omega_T = \Omega \times \{t = T\}.$$

[1] The references are given according to chapter; thus, for the first three chapters we refer to Volume I.

In Q, we consider the operator

(1.1) $$P = A\left(x, t, \frac{\partial}{\partial x}\right) + \frac{\partial}{\partial t} = A + \frac{\partial}{\partial t}$$

where A is given by

(1.2) $$A\,\varphi = \sum_{|p|,|q| \leq m} (-1)^{|p|} D_x^p(a_{pq}(x,t)\, D_x^q \varphi),$$

and D_x^p is defined in Chapter 2, Section 1.

The coefficients a_{pq} satisfy

(1.3) $$a_{pq} \in \mathscr{D}(\bar{Q}). \quad \square$$

Remark 1.2. The differentiability hypotheses (1.3) may be weakened in each of the following statements. However, in each statement, the "optimal" differentiability hypotheses are unknown; in fact, we recall – see Chapter 2, Remark 8.2 – that the systematic application of interpolation theory *cannot* yield optimal differentiability hypotheses. □

The boundary operators

We denote by B_j, $0 \leqq j \leqq m-1$, m *boundary operators* defined by

(1.4) $$B_j\,\varphi = \sum_{|h| \leqq m_j} b_{jh}\, D_x^h \varphi,$$

where the functions $b_{jh} = b_{jh}(x,t)$ satisfy

(1.5) $$b_{jh} \in \mathscr{D}(\bar{\Sigma})$$

(Remark 1.2 applies to the regularity hypotheses on b_{jh}).

We assume that

(1.6) $$\begin{cases} 0 \leqq m_j = \text{order of } B_j \leqq 2m-1; \\ \text{for each } t_0 \in [0,T], \text{ the system } \left\{B_j\left(x, t_0, \dfrac{\partial}{\partial x}\right)\right\}_{j=0}^{m-1} \\ \text{is normal on } \Gamma \text{ (see Chapter 2, Section 1.4)} \end{cases}$$

and that (*this is the fundamental hypothesis*):

$$\left\{\begin{array}{l} \text{for every } \theta \in \left[-\dfrac{\pi}{2}, \dfrac{\pi}{2}\right] \text{and } \forall t_0 \in (0,T] \\ \left\{A\left(x, t_0, \dfrac{\partial}{\partial x}\right) + (-1)^m e^{i\theta} D_y^{2m},\ B_j\left(x, t_0, \dfrac{\partial}{\partial x}\right)\right\} \\ \text{is a "regular elliptic" system in } \bar{\Omega} \times \mathbf{R}_y; \text{ in other words, for} \\ \text{every } \theta \in \left[-\dfrac{\pi}{2}, \dfrac{\pi}{2}\right] \text{and every } t_0 \in [0,T], \text{ the operator} \end{array}\right.$$

(1.7) $\left\{\begin{array}{l} A\left(x, t_0, \dfrac{\partial}{\partial x}\right) + (-1)^m e^{i\theta} D_y^{2m} \\ \text{is } \textit{properly elliptic in } \bar{\Omega} \times \mathbf{R}_y \text{ and the system of boundary operators} \\ \left\{B_j\left(x, t_0, \dfrac{\partial}{\partial x}\right)\right\}_{j=0}^{m-1} \\ \text{(which, according to hypothesis (1.6), is normal on } \Gamma \times \mathbf{R}_y\text{)} \\ \textit{covers} \text{ the operator} \\ A\left(x, t_0, \dfrac{\partial}{\partial x}\right) + (-1)^m e^{i\theta} D_y^{2m} \quad \text{on} \quad \Gamma \times \mathbf{R}_y. \end{array}\right.$

In (1.7) the *proper ellipticity* and *covering* hypotheses are to be understood in the sense of Chapter 2, Section 1 (in an obvious formulation for the case in question).

Hypotheses (1.6) and (1.7) are very general; for instance, we see immediately that if for every $t_0 \in [0, T]$ the operator $A\left(x, t_0, \frac{\partial}{\partial x}\right)$ is *strongly elliptic* in $\bar{\Omega}$ and if $B_j = \frac{\partial^j}{\partial \nu^j}$ (derivate of order j along the normal ν to Σ), $j = 0, \ldots, m-1$ (which corresponds to the conditions of Dirichlet on Σ), then (1.6) and (1.7) hold. ☐

1.2 Statement of the Problems

In this chapter we shall study – *always under the hypotheses of Section* 1.1 – the problems (mixed in the sense of J. Hadamard [1]):

(1.8) $$P u = A u + u' = f \quad \text{in} \quad Q\left(u' = \frac{\partial u}{\partial t}\right),$$

(1.9) $$B_j u = g_j \quad \text{on} \quad \Sigma, \quad 0 \leqq j \leqq m-1,$$

(1.10) $$u(x, 0) = u_0 \quad \text{on} \quad \Omega_0,$$

where the "functions" f, g_j and u_0 belong to various *Hilbert* classes. ☐

1.3 (Formal) Green's Formulas

The following Green's formulas should be compared to those in Chapter 2, Section 2.

The functions in the following formulas are always assumed to be "sufficiently regular"; *one* of the problems of the present chapter is in fact to *justify these formulas under "weak" hypotheses.* ☐

$\{B_j\}_{j=0}^{m-1}$ is completed by a system $\{S_j\}_{j=0}^{m-1}$ – which is not unique – in such a way as to make the system $\{B_j, S_j\}_{j=0}^{m-1}$ "normal" and "of Dirichlet" on Γ for each t (Chapter 2, Section 1). Then there exist $2m$ boundary operators C_j, T_j, $j = 0, \ldots, m-1$, with the following properties:

$$\text{the coefficients of } C_j,\ T_j \text{ belong to } \mathscr{D}(\overline{\Sigma}),$$

$$\text{order } C_j + \text{order } S_j = \text{order } B_j + \text{order } T_j = 2m - 1,$$

$$(1.11)\quad (A\,\varphi, \psi) - (\varphi, A^*\,\psi) = \sum_{j=0}^{m-1} \int_\Sigma S_j\,\varphi\,\overline{C_j\,\psi}\,d\Sigma - \sum_{j=0}^{m-1} \int_\Sigma B_j\,\varphi\,\overline{T_j\,\psi}\,d\Sigma$$

where φ and ψ are "regular" functions in Q, which, for example, belong to $\mathscr{D}(\overline{Q})$, where $(\ ,\ )$ denotes the scalar product in $L^2(Q)$ and where

$$A^*\,\psi = \sum_{|p|,|q| \leq m} (-1)^{|p|}\, D_x^p (\overline{a_{qp}(x,t)}\, D_x^q \psi).$$

The adjoint (in the sense of distributions on Q) of P is

$$P^* = A^* - \frac{\partial}{\partial t}$$

We have

$$(1.12)\quad \begin{cases} (P\,\varphi, \psi) - (\varphi, P^*\,\psi) = \sum_{j=0}^{m-1} \int_\Sigma S_j\,\varphi\,\overline{C_j\,\psi}\,d\Sigma - \\ \quad - \sum_{j=0}^{m-1} \int_\Sigma B_j\,\varphi\,\overline{T_j\,\psi}\,d\,\Sigma + \int_{\Omega_T} \varphi(x,T)\,\overline{\psi(x,T)}\,dx - \\ \quad - \int_{\Omega_0} \varphi(x,0)\,\overline{\psi(x,0)}\,dx. \quad \square \end{cases}$$

In order to simplify the notation, we set:

$$\vec{B}\,\varphi = \{B_j\,\varphi\}_{j=0}^{m-1}, \qquad \vec{C}\,\varphi = \{C_j\,\psi\}_{j=0}^{m-1},$$
$$\vec{S}\,\varphi = \{S_j\,\psi\}_{j=0}^{m-1}, \qquad \vec{T}\,\psi = \{T_j\,\psi\}_{j=0}^{m-1},$$

and we write (1.11) and (1.12) respectively in the form:

$$(1.11\text{a})\quad (A\,\varphi, \psi) - (\varphi, A^*\,\psi) = (\vec{S}\,\varphi, \vec{C}\,\psi)_\Sigma - (\vec{B}\,\varphi, \vec{T}\,\psi)_\Sigma,$$

$$(1.12\text{a})\quad (P\,\varphi, \psi) - (\varphi, P^*\,\psi) = (\vec{S}\,\varphi, \vec{C}\,\psi)_\Sigma - (\vec{B}\,\varphi, \vec{T}\,\psi)_\Sigma + \\ + (\varphi(T), \psi(T))_\Omega - (\varphi(0), \psi(0))_\Omega,$$

where the index Σ (resp. Ω) on the parentheses means integration on Σ (resp. Ω) and where for example $\varphi(T)$ denotes the function

$$x \to \varphi(x, T). \quad \square$$

1.4 First Existence and Uniqueness. Theorem (Statement)

Our first aim in this chapter is to prove the following theorem (the proof, as well as some additional results, are given in Sections 2 through 5 below):

Theorem 1.1. *Let the hypotheses of Section* 1.1 *be satisfied. Let f be given in $L^2(Q) = H^0(Q)$. Then there exists one and only one function u such that*

$$Pu = Au + u' = f \quad in \quad Q\left(u' = \frac{\partial u}{\partial t}\right), \tag{1.13}$$

$$B_j u = 0, \quad 0 \leqq j \leqq m-1, \tag{1.14}$$

$$u(x, 0) = 0, \tag{1.15}$$

$$u \in L^2(0, T; H^{2m}(\Omega)), \tag{1.16}$$

$$u' \in L^2(0, T; H^0(\Omega)), \tag{1.17}$$

(*we summarize* (1.16), (1.17); *by*: $u \in H^{2m,1}(Q)$ *see Section* 2 *below*).

Remark 1.3. In comparison with problem (1.9), (1.10), we have taken the functions g_j and u_0 to be zero. *Later on*, we shall consider the case of non-zero g_j and u_0. □

Remark 1.4. The space $L^2(0, T; H^{2m}(\Omega))$ (defined as a particular case of $L^2(0, T; X)$, see Chapter 1, Section 1) coincides with the space of functions v such that

$$v \in L^2(0, T; H^0(\Omega)) = L^2(Q) = H^0(Q),$$

$$D_x^p v \in H^0(Q) \quad \text{for} \quad |p| \leqq 2m. \ \square$$

Remark 1.5. In (1.7), $u' = \dfrac{\partial u}{\partial t}$ is defined as in Proposition 1.1, Chapter 1. □

Remark 1.6. Theorem 1.1 may be stated in the form of an *isomorphism* between two Hilbert spaces. We shall do this, in a more general setting, in Section 6. In any case, there exists a constant c such that

$$\|u\|_{L^2(0,T;H^{2m}(\Omega))} + \|u'\|_{L^2(0,T;H^0(\Omega))} \leqq c\, \|f\|_{L^2(Q)}. \ \square$$

1.5 Orientation

The proof of Theorem 1.1 (with numerous additional results) will be given in Sections 3, 4 and 5.

For the sake of convenience we start with a brief study of certain "Sobolev type" spaces which have a more complicated structure than those of Chapter 1; this is the aim of Section 2.

In Section 3 we shall make some general remarks about the Laplace transform and evolution equations; these remarks may be useful in other contexts as well.

In Section 4 we shall examine the case where A has coefficients independent of t. In Section 5 we shall obtain a regularity result via the method of Section 4. Finally in Section 6 we shall examine the general case of coefficients which depend on x *and* t.

2. The Spaces $H^{r,s}(Q)$. Trace Theorems. Compatibility Relations

2.1 $H^{r,s}$-Spaces

Let r and s be two non-negative real numbers. For Ω an open set in $\mathbf{R}^n$, we define:

(2.1) $$H^{r,s}(Q) = H^0(0,T;H^r(\Omega)) \cap H^s(0,T;H^0(\Omega)), \quad (Q = \Omega \times]0,T[),$$

which is a Hilbert space with the norm

$$\left(\int_0^T \|u(t)\|^2_{H^r(\Omega)}\, dt + \|u\|^2_{H^s(0,T;H^0(\Omega))}\right)^{1/2}.$$

In (2.1), $H^0(0, T; H^r(\Omega)) = L^2(0, T; H^r(\Omega))$, where $H^r(\Omega)$ is defined in Chapter 1, Section 9, and $H^s(0, T; H^0(\Omega))$ is defined in the same way as $H^s(0, T)$; see Chapter 1, Section 9; for example, if X is a Hilbert space,

$$H^s(0, T; X) = [H^m(0, T; X), H^0(0, T; X)]_\theta, \quad (1 - \theta)\, m = s,$$

integer $m > s$, and

$$H^m(0, T; X) = \{v \mid v, v', \ldots, v^{(m)} \in L^2(0, T; X)\}. \quad \square$$

Remark 2.1. The space $H^{r,s}(Q)$ is the space of *restrictions to* $Q = \Omega \times]0, T[$ of functions in $H^{r,s}(\Omega \times \mathbf{R})$; of course, according to (2.1):

$$H^{r,s}(\Omega \times \mathbf{R}) = H^0(\mathbf{R}; H^r(\Omega)) \cap H^s(\mathbf{R}; H^0(\Omega))$$

and this time we may define $H^s(\mathbf{R}; H^0(\Omega))$ by a *Fourier transform in* t:

$$v \in H^s(\mathbf{R}; H^0(\Omega)) \Leftrightarrow (1 + |\tau|^2)^{s/2}\, \hat{v} \in L^2(\mathbf{R}_\tau; H^0(\Omega)) = L^2(\Omega \times \mathbf{R}_\tau)$$

if $\hat{v}(\tau) =$ Fourier transform in t of $v(t)$. $\square$

Remark 2.2. We have assumed that Ω is a bounded open set in $\mathbf{R}^n$, whose boundary Γ is an $(n-1)$-dimensional, infinitely differentiable variety such that Ω is locally on only one side of Γ (see Section 1.1).

Then

(2.2) $H^{r,s}(Q)$ = space of restrictions to Q of functions

in $H^{r,s}(\mathbf{R}_x^n \times \mathbf{R}_t)$

and when Ω and $]0, T[$ are replaced by $\mathbf{R}_x^n$ and $\mathbf{R}_t$, definition (2.1) is *equivalent* to (see Chapter 1, Section 13.2)

(2.3) $H^{r,s}(\mathbf{R}_x^n \times \mathbf{R}_t) = \{v \mid [(1 + |\xi|^2)^{r/2} + (1 + |\tau|^2)^{s/2}]\, \hat{v} \in L^2(\mathbf{R}_\xi^n \times \mathbf{R}_\tau)\}$

where $\hat{v}(\xi, \tau)$ = Fourier transform in x and t of $v(x, t)$.

Under these hypotheses, *$\mathscr{D}(\bar{Q})$ is dense in $H^{r,s}(Q)$* (proof analogous to the density theorems of Chapter 1, Sections 8 and 9). □

An interpolation property:

Proposition 2.1. *For $r, s \geqq 0$ and $\theta \in]0, 1[$, we have*

(2.4) $$[H^{r,s}(Q), L^2(Q)]_\theta = H^{(1-\theta)r,(1-\theta)s}(Q).$$

More generally, if $r, s, \varrho, \sigma \geqq 0$, we have:

(2.5) $$[H^{r,s}(Q), H^{\varrho,\sigma}(Q)]_\theta = H^{(1-\theta)r+\theta\varrho,(1-\theta)s+\theta\sigma}(Q).$$

Proof. We restrict the proof to (2.4); the proof of (2.5) follows the same principle.

1) We shall use Chapter 1, Section 13.1. We note: $L^2(0,T; H^r(\Omega))$ is the domain in $L^2(Q)$ of a positive self-adjoint operator R (with norm equivalent to the norm of the graph) and $H^s(0, T; H^0(\Omega))$ is the domain in $L^2(Q)$ of a positive self-adjoint operator S — and the operators R and S *commute* (one is an operator in x, independent of t and the other an operator in t, independent of x). Then

$$H^{r,s}(Q) = D(R) \cap D(S)$$

and, according to Chapter 1, Section 13.1, we therefore have

$$[H^{r,s}(Q), H^0(Q)]_\theta = X \cap Y,$$

where

$$X = [L^2(0, T; H^r(\Omega)) = D(R), H^0(Q)]_\theta$$

and

$$Y = [H^s(0, T; H^0(\Omega)) = D(S), H^0(Q)]_\theta.$$

2) Then (2.4) follows from the two identities:

(2.6) $$[L^2(0, T; H^r(\Omega)), L^2(0, T; H^0(\Omega))]_\theta = L^2(0, T; H^{(1-\theta)r}(\Omega))$$

and

$$(2.7)\qquad [H^s(0,T;H^0(\Omega)), L^2(0,T;H^0(\Omega))]_\theta = H^{(1-\theta)s}(0,T;H^0(\Omega)).$$

Identity (2.6) is a particular case of the general identity (C_0 and C_1 are Hilbert spaces):

$$[L^2(0,T;C_0), L^2(0,T;C_1)]_\theta = L^2(0,T;[C_0, C_1]_\theta)$$

(see Chapter 1, Section 14.2)

and of

$$[H^r(\Omega), H^0(\Omega)]_\theta = H^{(1-\theta)r}(\Omega)$$

(Chapter 1, Section 9).

Identity (2.7) is the analogue (with values in a Hilbert space) of Chapter 1, Section 9. □

Some subspaces of $H^{r,s}(Q)$.

We shall write:

$$(2.8)\qquad H_{0,}^{r,s}(Q) = L^2(0,T;H_0^r(\Omega)) \cap H^s(0,T;H^0(Q));$$

and it can be verified that this space coincides with the closure in $H^{r,s}(Q)$ of the subspace of functions which vanish in a (variable) neighborhood of $\Sigma = \Gamma \times]0,T[$;

$$(2.9)\qquad H_{,0}^{r,s}(Q) = L^2(0,T;H^r(\Omega)) \cap H_0^s(0,T;H^0(\Omega)),$$

which is the closure in $H^{r,s}(Q)$ of the subspace of functions which vanish in the neighborhood of $t = 0$ and of $t = T$;

$$(2.10)\qquad H_{0,0}^{r,s}(Q) = H_{0,}^{r,s}(Q) \cap H_{,0}^{r,s}(Q),$$

which is the closure of $\mathscr{D}(Q)$ in $H^{r,s}(Q)$. □

The space $H^{r,s}(\Sigma)$.

Definition (2.1) extends without difficulty to the case where Ω is replaced by its boundary Σ; it suffices to use the spaces $H^r(\Gamma)$ instead of $H^r(\Omega)$ (see Chapter 1, Section 7); we obtain:

$$(2.11)\qquad H^{r,s}(\Sigma) = H^0(0,T;H^r(\Gamma)) \cap H^s(0,T;H^0(\Gamma)).$$

A result analogous to Proposition 2.1 holds for these spaces. We recall that we have already introduced these spaces in Chapter 1, Section 13.3 with a definition equivalent to (2.11).

Since $H_0^r(\Gamma) = H^r(\Gamma)$ ($\mathscr{D}(\Gamma)$ is dense in $H^r(\Gamma)$), there is no distinction between $H_{0,}^{r,s}(\Sigma)$ and $H^{r,s}(\Sigma)$; but we may set (by analogy with (2.9)):

$$(2.12)\qquad H_{,0}^{r,s}(\Sigma) = H^0(0,T;H^r(\Gamma)) \cap H_0^s(0,T;H^0(\Gamma)). \quad \square$$

2.2 First Trace Theorem

Theorem 2.1. *For* $u \in H^{r,s}(Q)$ *with* $r > \frac{1}{2}$, $s \geqq 0$, *we may define:*

$$(2.13)\quad \begin{cases} \dfrac{\partial^j u}{\partial \nu^j} \quad \text{on } \Sigma \text{ if } j < r - \dfrac{1}{2} \quad (\text{integer } j \geqq 0), \\ \dfrac{\partial^j u}{\partial \nu^j} \in H^{\mu_j, \nu_j}(\Sigma) \\ \left(\dfrac{\partial}{\partial \nu} = \begin{array}{l}\text{normal derivative on } \Sigma, \\ \text{oriented toward the interior of } Q\end{array}\right) \end{cases}$$

where

$$(2.14)\quad \frac{\mu_j}{r} = \frac{\nu_j}{s} = \frac{r - j - \frac{1}{2}}{r} \quad (\nu_j = 0 \text{ if } s = 0),$$

and where $u \to \dfrac{\partial^j u}{\partial \nu^j}$ *are continuous linear mappings of*

$$H^{r,s}(Q) \to H^{\mu_j, \nu_j}(\Sigma)\ ^{((1))}.$$

We may also define, for $s > \frac{1}{2}$, $r \geqq 0$:

$$(2.15)\quad \begin{cases} \dfrac{\partial^k u}{\partial t^k}(x, 0) \quad \text{on } \Omega \text{ if } k < s - \frac{1}{2} \quad (\text{integer } k \geqq 0), \\ \dfrac{\partial^k u}{\partial t^k}(x, 0) \in H^{p_k}(\Omega), \end{cases}$$

where

$$(2.16)\quad p_k = \frac{r}{s}\left(s - k - \frac{1}{2}\right),$$

and where $u \to \dfrac{\partial^k u}{\partial t^k}(x, 0)$ *are continuous mappings of* $H^{r,s}(Q) \to$ $\to H^{p_k}(\Omega)$ ((1)).

Before proving the theorem, let us point out that the mapping

$$(2.17)\quad u \to \left\{\frac{\partial^j u}{\partial \nu^j}, \frac{\partial^k u}{\partial t^k}(x, 0)\right\}, \quad j < s - \frac{1}{2}, k < r - \frac{1}{2},$$

of

$$H^{r,s}(Q) \to \prod_j H^{\mu_j, \nu_j}(\Sigma) \times \prod_k H^{p_k}(\Omega)$$

is not surjective (see the sections immediately following this one).

Proof of Theorem 2.1. 1) *Proof of* (2.13) *and* (2.14). All spaces under consideration are *local* (see Chapter 1, Section 7); we may therefore

((1)) The traces are always understood by extension by continuity from $\mathscr{D}(\bar{Q})$, wich is dense in $H^{r,s}(Q)$ (see Remark 2.2).

reduce the problem to the case

$$\Omega = \{x \mid x_n > 0\}.$$

Let

$$x = \{x', x_n\}, \quad x' = \{x_1, \ldots, x_{n-1}\}.$$

Then $u \in H^{r,s}(\Omega \times]0, T[)$ is *equivalent* to:

$$u \in H^r(\mathbf{R}_{+,x_n}; H^0(\mathbf{R}^{n-1}_{x'} \times]0, T[)),$$

$$u \in H^0(\mathbf{R}_{+,x_n}; H^{r,0}_{x',t}),$$

$$u \in H^0(\mathbf{R}_{+,x_n}; H^{0,s}_{x',t}),$$

or

$$u \in H^0(]0, \infty[; H^{r,s}_{x',t}) \cap H^r(]0, \infty[; H^{0,0}_{x',t}) \tag{2.18}$$

(we have set: $H^{r,s}_{x,t} = H^0(]0, T[; H^r(\mathbf{R}^{n-1}_{x'})) \cap H^s(]0, T[; H^0(\mathbf{R}^{n-1}_{x'})))$.

Now we apply Theorem 4.2 of Chapter 1, with

$$X = H^{r,s}_{x',t}, \quad Y = H^{0,0}_{x',t} \tag{2.19}$$

(and r instead of s; Theorem 4.2 remains unchanged if $\mathbf{R}$ is replaced by the half-line $]0, \infty[$); we obtain:

$$u^{(j)}(0) = \frac{\partial^j}{\partial x_n^j} u(x', 0, t) \in [X, Y]_{\frac{j+1/2}{r}}$$

which, together with (2.19) and Proposition 2.1, yields (2.13) and (2.14); indeed

$$[X, Y]_{\frac{j+1/2}{r}} = H^{\left(1 - \frac{j+1/2}{r}\right)r, \left(1 - \frac{j+1/2}{r}\right)s}_{x',t}$$

whence (2.14).

2) *Proof of* (2.15) *and* (2.16). This is an immediate consequence of Theorem 4.2 of Chapter 1: since

$$u \in H^0(0, T; H^r(\Omega)) \cap H^s(0, T; H^0(\Omega)),$$

we apply Theorem 4.2 (having replaced $\mathbf{R}$ – which is permissible – with $]0, T[$) with $X = H^r(\Omega)$, $Y = H^0(\Omega)$; then

$$u^{(k)}(0) = \frac{\partial^k u}{\partial t^k}(x, 0) \in [X, Y]_{\frac{j+1/2}{s}} = H^{r\left(1 - \frac{j+1/2}{s}\right)}(\Omega)$$

(according to Chapter 1, Section 9). □

2.3 Local Compatibility Relations

We shall see that the mapping (2.17) *is not surjective*; the functions $\dfrac{\partial^j u}{\partial \nu^j}$ and $\dfrac{\partial^k u}{\partial t^k}(x, 0)$ satisfy additional relations, called *compatibility relations* ($\mathcal{R}.\mathcal{C}.$).

Formally, it is clear that

$$(2.20)\qquad \frac{\partial^k}{\partial t^k}\left(\frac{\partial^j u}{\partial \nu^j_\cdot}\bigg|_\Sigma\right)_{t=0} = \gamma_j\left(\frac{\partial^k u}{\partial t^k}(x,0)\right),$$

(γ_j = normal derivative on Γ, of order j).

The $(\mathscr{R}.\mathscr{C}.)$ express (2.20) – whenever this has meaning – with a "limit case" which will be investigated in the following sections. We shall prove

Proposition 2.2. *Let* $r, s > 0$ *with* $1 - \frac{1}{2}\left(\frac{1}{r}+\frac{1}{s}\right) > 0$ ((1)); *relations* (2.20) *are valid for all integer couples* j, k *such that*

$$(2.21)\qquad \frac{j}{r}+\frac{k}{s} < 1 - \frac{1}{2}\left(\frac{1}{r}+\frac{1}{s}\right).$$

Proof. Again we reduce the problem to the case $\Omega = \{x \mid x_n > 0\}$. We have to show that

$$(2.22)\qquad \frac{\partial^k}{\partial t^k}\frac{\partial^j u}{\partial x_n^j}(x',0,0) = \frac{\partial^j}{\partial x_n^j}\frac{\partial^k u}{\partial t^k}(x',0,0).$$

According to (2.13):

$$\frac{\partial^j u}{\partial x_n^j}(x',0,t) \in H^{\mu_j,\nu_j}(\mathbf{R}^{n-1}_{x'} \times]0,T[)\ ^{((2))}$$

and again applying Theorem 2.1 (but this time (2.15)), we have:

$$\frac{\partial^k}{\partial t^k}\frac{\partial^j u}{\partial x_n^j}(x',0,0) \in H^{\frac{\mu_j}{\nu_j}(\nu_j-k-1/2)}(\mathbf{R}^{n-1}_{x'})$$

if $k < \nu_j - \frac{1}{2}$, i.e. (2.21) (applying (2.14)) and

$$(2.23)\qquad u \to \frac{\partial^k}{\partial t^k}\frac{\partial^j u}{\partial x_n^j}(x',0,0)$$

is a continuous mapping of

$$H^{r,s}(Q) \to H^{\frac{\mu_j}{\nu_j}(\nu_j-k-1/2)}(\mathbf{R}^{n-1}_{x'}).$$

Similarly, according to (2.15),

$$\frac{\partial^k u}{\partial t^k}(x,0) \in H^{p_k}(\Omega),$$

((1)) Note that in this case; $r > \frac{1}{2}$, $s > \frac{1}{2}$.

((2)) Note that in this case: $\mu_j > \frac{1}{2}$, $\nu_j > \frac{1}{2}$.

therefore

$$\gamma_j \frac{\partial^k u}{\partial t^k}(x, 0) \in H^{p_k - j - 1/2}(\Gamma) \quad \text{(Chapter 1, Section 9)}$$

$$\text{if } j < p_k - \tfrac{1}{2},$$

i.e. (2.21) $\left(\text{and } p_k - j - \dfrac{1}{2} = \dfrac{\mu_j}{\nu_j}\left(\nu_j - k - \dfrac{1}{2}\right) \text{ is satisfied}\right)$.

Furthermore

$$u \to \frac{\partial^j}{\partial x_n^j} \frac{\partial^k u}{\partial t^k}(x', 0, 0) \tag{2.24}$$

is a continuous mapping of

$$H^{r,s}(Q) \to H^{\frac{\mu_j}{\nu_j}(\nu_j - k - 1/2)}(\mathbf{R}_{x'}^{n-1}).$$

Since the mappings (2.23) and (2.24) coincide on the *dense* subspace $\mathscr{D}(\bar{\Omega})$, the result follows. □

Remark 2.3. In the proof of Proposition 2.2 we may also use the following result:

Proposition 2.3. *If* $u \in H^{r,s}(Q)$, r *and* $s > 0$, j, k *integers such that* $1 - \left(\dfrac{j}{r} + \dfrac{k}{s}\right) \geqq 0$, *we have*

(2.25)

$$D_x^j \frac{\partial^k}{\partial t^k} u \in H^{\mu,\nu}(Q) \qquad (D_x^j = j\text{th order derivative with respect to } x),$$

with

$$\frac{\mu}{r} = \frac{\nu}{s} = 1 - \left(\frac{j}{r} + \frac{k}{s}\right). \tag{2.26}$$

Proof. We reduce the problem to the case $Q = \mathbf{R}_x^n \times \mathbf{R}_t$ and use the Fourier transform in x and t. Then the result follows from

$$|\xi|^j |\tau|^k [(1 + |\xi|^2)^{\mu/2} + (1 + \tau^2)^{\nu/2}] \leqq c[(1 + |\xi|^2)^{r/2} + (1 + \tau^2)^{s/2}] \quad \square$$

State of the problem

We shall now see that when the *equality* holds in (2.21) we again have a compatibility relation, but *this time a non-local one*. The relations are due to P. Grisvard [4] and [8]; we shall specify them first for a (fundamental) particular case. This is the aim of the next section.

2.4 Global Compatibility Relations for a Particular Case

We consider a Hilbert space X and the space

$$H^{r,s}(\mathbf{R}_+^2; X), \quad \mathbf{R}_+ =]0, \infty[, \quad \{x, t\} \in \mathbf{R}_+^2.$$

Theorem 2.2. *Assume that*

$$\frac{1}{r} + \frac{1}{s} = 2, \quad r, s > 0. \tag{2.27}$$

Then (since $r > \frac{1}{2}$, $s > \frac{1}{2}$) we may define

$$\begin{cases} u(x, 0) = f(x) \in H^{1/2}(\mathbf{R}_+; X)^{((1))} \\ u(0, t) = g(t) \in H^{1/2}(\mathbf{R}_+; X). \end{cases} \tag{2.28}$$

We have (global compatibility relation):

$$\int_0^\infty \| f(\sigma^s) - g(\sigma^r) \|_X^2 \frac{d\sigma}{\sigma} < \infty. \tag{2.29}$$

Proof. 1) To u, we associate the function $v = v(x, t)$ (for reasons which will become clear later on) defined by

$$v(x, t) = \sum_{k=1}^{m} \sum_{l=1}^{n} \alpha_k \beta_l u(k x, l t), \tag{2.30}$$

where

$$\begin{cases} m = [r] + 1, \quad n = [s] + 1 \quad ([\sigma] = \text{integer part of } \sigma) \\ \sum_k \alpha_k = \sum_l \beta_l = 1, \\ \sum_k k^i \alpha_k = 0, \quad i = 1, 2, \ldots, m - 1, \\ \sum_l l^j \beta_l = 0, \quad j = 1, \ldots, n - 1. \end{cases} \tag{2.31}$$

The function v also belongs to $H^{r,s}(\mathbf{R}_+^2; X)$ and

$$\begin{cases} v(x, 0) = \sum_k \alpha_k f(k x) = \varphi(x), \\ v(0, t) = \sum_l \beta_l g(l t) = \psi(t). \end{cases} \tag{2.32}$$

But the linear mapping

$$f \to \varphi - f \tag{2.33}$$

is a continuous mapping of

$$H^1(\mathbf{R}_+; X) \to H_0^1(\mathbf{R}_+; X) \qquad (\text{since} \sum_k \alpha_k = 1)$$

and of

$$H^0(\mathbf{R}_+; X) \to H^0(\mathbf{R}_+; X);$$

((1)) Apply Theorem 2.1 (valid without change for functions taking their values in a Hilbert space.)

therefore by interpolation and application of the results of Chapter 1, Sections 9 and 11.5, Theorem 11.7 (valid without change for functions with values in a Hilbert space):

$$(2.34)\quad \begin{cases} (2.33) \text{ is a continuous linear mapping of} \\ H^{1/2}(\mathbf{R}_+ ; X) \to H_{00}^{1/2}(\mathbf{R}_+ ; X), \end{cases}$$

where

$$(2.35)\quad H_{00}^{1/2}(\mathbf{R}_+ ; X) = \left\{ v \mid v \in H^{1/2}(\mathbf{R}_+ ; X), \int_0^\infty \frac{1}{\sigma} \| v(\sigma) \|_X^2 \, d\sigma < \infty \right\}.$$

The same remarks hold for the mapping $g \to \psi - g$.

Therefore, in the notation of (2.28) and (2.32):

$$\int_0^\infty \| f(\sigma) - \varphi(\sigma) \|_X^2 \frac{d\sigma}{\sigma} < \infty, \quad \int_0^\infty \| g(\sigma) - \psi(\sigma) \|_X^2 \frac{d\sigma}{\sigma} < \infty$$

or, equivalently

$$\int_0^\infty \| f(\sigma^s) - \varphi(\sigma^s) \|_X^2 \frac{d\sigma}{\sigma} < \infty, \quad \int_0^\infty \| g(\sigma^r) - \psi(\sigma^r) \|_X^2 \frac{d\sigma}{\sigma} < \infty.$$

Therefore we shall have obtained (2.29) if we can show:

$$(2.36)\quad \int_0^\infty \| \varphi(\sigma^s) - \psi(\sigma^r) \|_X^2 \frac{d\sigma}{\sigma} < \infty.$$

2) According to (2.32), we may write

$$\varphi(\sigma^s) - \psi(\sigma^r) = v(\sigma^s, 0) - v(\sigma^s, \sigma^r) + v(\sigma^s, \sigma^r) - v(0, \sigma^r)$$

and since the variables x and t play analogous roles, we shall have (2.36) if we can show that

$$(2.37)\quad \int_0^\infty \| v(\sigma^s, 0) - v(\sigma^s, \sigma^r) \|_X^2 \frac{d\sigma}{\sigma} < \infty.$$

We shall consider the two cases $s \leqq 1$ and $s > 1$.

3) *The case* $s \leqq 1$.

We shall need the following lemma (which we shall prove further on):

Lemma 2.1. *Let* $\frac{1}{2} < s < \frac{3}{2}$. *Then, if* $u \in H^s(\mathbf{R}_+ ; X)$, *there exists a constant* c *such that*

$$(2.38)\quad \| u(\xi + h) - u(\xi) \|_X \leqq c\, h^{s-1/2}, \quad \xi, \xi + h \in \mathbf{R}_+.$$

Since $v \in H^{r,s}(\mathbf{R}_+^2; X)$, we have:

$$v \in L^2(\mathbf{R}_+; H^s(\mathbf{R}_+; X)),$$

and therefore, according to (2.38),

$$v \in L^2(\mathbf{R}_+; \mathrm{Lip}_{s-1/2}(\mathbf{R}_+; X))^{((1))},$$

therefore

$$\| v(x, t) - v(x, t') \|_X \leqq w(x) \, |t - t'|^{s-1/2}, \qquad w \in L^2(\mathbf{R}_+).$$

Then

$$\| v(\sigma^s, 0) - v(\sigma^s, \sigma^r) \|_X \leqq w(\sigma^s) \, \sigma^{r(s-1/2)}$$

$$= w(\sigma^s) \, \sigma^{s/2} \quad \text{(according to (2.27)]}.$$

and

$$\int_0^\infty \| v(\sigma^s, 0) - v(\sigma^s, \sigma^r) \|_X^2 \frac{d\sigma}{\sigma} = \int_0^\infty w(\sigma)^2 \, d\sigma < +\infty.$$

4) *The case* $s > 1$.

Thanks to (2.31), we have

$$\frac{\partial^j v}{\partial t^j}(x, 0) = 0, \quad j = 1, \ldots, [s] - 1 \tag{2.39}$$

and therefore, setting $\mu = [s] = n - 1$:

$$v(x, t) - v(x, 0) = \frac{1}{(\mu - 1)!} \int_0^t (t - \sigma)^{\mu - 1} \frac{\partial^\mu v}{\partial t^\mu}(x, \sigma) \, d\sigma. \tag{2.40}$$

Consequently

$$\| v(x, t) - v(x, 0) \|_X \leqq c \, t^{\mu - 1} \int_0^t \left\| \frac{\partial^\mu v}{\partial t^\mu}(x, \sigma) \right\|_X d\sigma, \tag{2.41}$$

where c denotes a constant.

But according to (2.30):

$$\frac{\partial^\mu v}{\partial t^\mu}(x, t) = \sum_l l^\mu \beta_l \, h(x, l\,t),$$

where

$$h(x, t) = \sum_k \alpha_k \frac{\partial^\mu u}{\partial t^\mu}(k\,x, t).$$

If $\beta_l' = l^\mu \beta_l$ and if we set

$$\chi(x, t) = \sum_l \beta_l' \, h(x, l\,t),$$

[((1))] $\mathrm{Lip}_\theta(\mathbf{R}_+; X)$ = space of continuous functions f, mapping $t \geqq 0 \to X$, such that $\sup\limits_{t,h>0} h^{-\theta} \, \| f(t+h) - f(t) \|_X < \infty$.

we have

$$(2.42)\qquad h - \chi = -\frac{\partial^\mu v}{\partial t^\mu}.$$

But

$$h \in L^2(\mathbf{R}_{+,x}; H^{s-\mu}(\mathbf{R}_+; X))$$

and since $\sum_i \beta_i' = 1$, we see (as for (2.34), (2.35)) that

$$h - \chi \in L^2(\mathbf{R}_{+,x}; H_0^{s-\mu}(\mathbf{R}_+; X));$$

applying Chapter 1, Section 11 and (2.42), it follows that

$$(2.43)\qquad \int_0^\infty dx \int_0^\infty t^{-2(s-\mu)} \left\| \frac{\partial^\mu v}{\partial t^\mu}(x, t) \right\|_X^2 dt < \infty.$$

But then

$$\int_0^\infty \| v(\sigma^s, 0) - v(\sigma^s, \sigma^r) \|_X^2 \frac{d\sigma}{\sigma} = \int_0^\infty \| v(\sigma, 0) - v(\sigma, \sigma^{r/s}) \|_X^2 \frac{d\sigma}{\sigma} \leqq$$

$$\leqq \text{(according to (2.41))} \leqq c \int_0^\infty \sigma^{2\frac{r}{s}(\mu-1)} \left(\int_0^{\sigma^{r/s}} \left\| \frac{\partial^\mu v}{\partial t^\mu}(\sigma, y) \right\|_X dy \right)^2 \frac{d\sigma}{\sigma} \leqq$$

$$\leqq \int_0^\infty \sigma^{2\frac{r}{s}(\mu-1)} \left(\int_0^{\sigma^{r/s}} y^{-2(s-\mu)} \left\| \frac{\partial^\mu v}{\partial r^\mu}(\sigma, y) \right\|_X^2 dy \right) \left(\int_0^{\sigma^{r/s}} y^{2(s-\mu)} dy \right) \frac{d\sigma}{\sigma}$$

$$= c \int_0^\infty \left(\int_0^{\sigma^{r/s}} y^{-2(s-\mu)} \left\| \frac{\partial^\mu v}{\partial t^\mu}(\sigma, y) \right\|_X^2 dy \right) d\sigma \qquad \text{(applying (2.27)},$$

which is *finite* according to (2.43). □

Proof of Lemma 2.1. Of course it is equivalent to establish an analogous result for

$$u \in H^s(\mathbf{R}; X).$$

We apply the Fourier transform; if for example

$$\hat{u}(\xi) = \int_{-\infty}^{+\infty} e^{-2\pi i x \xi} u(x)\, dx,$$

we have:

$$\| u(x+h) - u(x) \|_X \leqq \int_{-\infty}^{+\infty} | e^{2\pi i x \xi} - 1 | \, \| \hat{u}(\xi) \|_X \, d\xi.$$

Choosing c_0 such that $|e^{it} - 1| \leqq 2|t|$ for $|t| \leqq c_0$, we obtain

$$\|u(x+h) - u(x)\|_X \leqq 4\pi|h| \int_{|\xi| \leqq c_0/2\pi|h|} \|\xi\hat{u}(\xi)\|_X d\xi +$$

$$+ 2 \int_{|\xi| > c_0/2\pi|h|} \|\hat{u}(\xi)\|_X d\xi \leqq 4\pi|h| \int_{|\xi| \leqq c_0/2\pi|h|} \||\xi|^s \hat{u}\|_X |\xi|^{1-s} d\xi +$$

$$+ 2 \int_{|\xi| > c_0/2\pi|h|} |\xi|^{-s} \||\xi|^s \hat{u}\|_X d\xi \leqq$$

$$\leqq 4\pi|h| \left(\int_{|\xi| \leqq c_0/2\pi|h|} |\xi|^{2(1-s)} d\xi \right)^{1/2} \left(\int_{\mathbf{R}} \||\xi|^s \hat{u}\|_X^2 d\xi \right)^{1/2} +$$

$$+ 2 \left(\int_{|\xi| \geqq c_0/2\pi|h|} |\xi|^{-2s} d\xi \right)^{1/2} \left(\int_{\mathbf{R}} \||\xi|^s \hat{u}\|_X^2 d\xi \right)^{1/2} \leqq c|h|^{s-1/2}. \quad \square$$

2.5 General Compatibility Relations

We are now ready to specify *all* compatibility relations satisfied by the traces of $u \in H^{r,s}(Q)$. With the aid of *local maps*, we can establish a correspondence between each $u \in H^{r,s}(Q)$ and a *finite number* of functions, denoted by

$$u_1, u_2, \ldots, u_N,$$

which define u uniquely and are such that

(2.44) $$u_i \in H^{r,s}(\mathbf{R}^n_+ \times]0, T[), \quad \mathbf{R}^n_+ = \{x \mid x_n > 0\}.$$

On the other hand, there exist compatibility relations *of the same kind* on the hypersurfaces $\{x \in \Gamma, t = 0\}$ *and* $\{x \in \Gamma, t = T\}$. *Therefore, we do not lose any generality by assuming* $T = +\infty$ and consequently *it is sufficient to specify all* ($\mathcal{R}.\mathcal{C}.$) *for the case* $\mathbf{R}^n_+ \times]0, \infty[$. Therefore, let

$$u \in H^{r,s}(\mathbf{R}^n_+ \times]0, \infty[), \quad r, s > 0, \quad 1 - \frac{1}{2}\left(\frac{1}{r} + \frac{1}{s}\right) > 0.$$

Set:

(2.45) $$\begin{cases} \dfrac{\partial^k u}{\partial t^k}(x, 0) = f_k(x), \\ \\ \dfrac{\partial^j u}{\partial x_n^j}(x', 0, t) = g_j(x', t). \end{cases}$$

According to Theorem 2.1, we have:

(2.46) $$\begin{cases} \{f_k, g_j\} \in F = \prod_{k < s - 1/2} H^{p_k}(\mathbf{R}^n_+) \times \prod_{j < r - 1/2} H^{\mu_j, \nu_j}(\Sigma), \\ \\ p_k, \mu_j, \nu_j \text{ given by (2.16) and (2.14).} \end{cases}$$

According to (2.20), we know that

$$\frac{\partial^k g_j}{\partial t^k}(x', 0) = \frac{\partial^j f_k}{\partial x_n^j}(x', 0) \quad \text{if} \quad \frac{j}{r} + \frac{k}{s} < 1 - \frac{1}{2}\left(\frac{1}{r} + \frac{1}{s}\right). \tag{2.47}$$

According to Theorem 2.2, we have:

$$\begin{cases} \int_0^\infty \int_{\mathbf{R}^{n-1}} \left| \frac{\partial^j f_k}{\partial x_n^j}(x', \sigma^s) - \frac{\partial^k g_j}{\partial t^k}(x', \sigma^r) \right|^2 dx' \frac{d\sigma}{\sigma} < \infty \\ \text{if} \quad \frac{j}{r} + \frac{k}{s} = 1 - \frac{1}{2}\left(\frac{1}{r} + \frac{1}{s}\right). \end{cases} \tag{2.48}$$

Indeed, according to Proposition 2.3,

$$\frac{\partial^{j+k}}{\partial x_n^j \partial t^k} u \in H^{\mu,\nu}(\mathbf{R}_+^n \times]0, \infty[)$$

and from (2.26) (with (2.48)) we deduce that

$$\frac{1}{\mu} + \frac{1}{\nu} = 2.$$

The function

$$x_n, t \to \text{“} x' \to u(x', x_n, t_n)\text{”}$$

thus belongs to $H^{\mu,\nu}(\mathbf{R}_+^2; H^0(\mathbf{R}_x^{n-1}))$ and therefore (2.48) is just the interpretation of (2.29). □

We denote by F_0 the vector subspace of the product space F of elements $\{f_k, g_j\}$ which satisfy (2.47) and (2.48); F_0 is a Hilbert space for the norm whose square is given by

$$\|\{f_k, g_j\}\|_F^2 + \sum_{k,j} \int_0^\infty \int_{\mathbf{R}^{n-1}} \left| \frac{\partial^j f_k}{\partial x_n^j}(x', \sigma^s) - \frac{\partial^k g_j}{\partial t^k}(x', \sigma^r) \right|^2 dx' \frac{d\sigma}{\sigma}.$$

Then the fundamental result is:

Theorem 2.3. $u \to \{f_k, g_j\}$ *(defined by (2.45)) is a continuous linear surjection of*

$$H^{r,s}(\mathbf{R}_+^n \times]0, \infty[) \to F_0, \quad \textit{if} \quad r > 0, \quad s > 0, \quad 1 - \frac{1}{2}\left(\frac{1}{r} + \frac{1}{s}\right) > 0.$$

Proof. Only the *surjectivity* remains to be shown.

According to Theorem 4.2 of Chapter 1 (applied as in 1) of the proof of Theorem 2.1) there exists a function $v \in H^{r,s}(\mathbf{R}^n_+ \times]0, \infty[)$ such that

$$\frac{\partial^j v}{\partial x_n^j}(x', 0, t) = g_j(x', t), \quad 0 \leqq j < r - \frac{1}{2}. \tag{2.49}$$

It remains to construct $w \in H^{r,s}(\mathbf{R}^n_+ \times]0, \infty[)$ such that

$$\begin{cases} \dfrac{\partial^j w}{\partial x_n^j}(x', 0, t) = 0, \\ \dfrac{\partial^k w}{\partial t_k}(x, 0) = f_k - h_k, \quad 0 \leqq k < s - \dfrac{1}{2}, \end{cases} \tag{2.50}$$

where

$$h_k(x, 0) = \frac{\partial^k v}{\partial t^k}(x, 0). \tag{2.51}$$

But since by hypothesis $\{f_k, g_j\} \in F_0$ and since $\{h_k, g_j\} \in F_0$, we obtain:

$$\{f_k - h_k, 0\} \in F_0.$$

Thus, if we set

$$\varphi_k = f_k - h_k \in H^{p_k}(\mathbf{R}^n_+), \tag{2.52}$$

we have $\{\varphi_k, 0\} \in F_0$, that is

$$\frac{\partial^j \varphi_k}{\partial x_n^j}(x', 0) = 0 \quad \text{if} \quad j < p_k - \frac{1}{2} \tag{2.53}$$

and

$$\int_0^\infty \int_{\mathbf{R}^{n-1}} \left| \frac{\partial^j \varphi_k}{\partial x_n^j}(x', \sigma) \right|^2 dx' \frac{d\sigma}{\sigma} < \infty \quad \text{if} \quad j = p_k - \frac{1}{2}. \tag{2.54}$$

Therefore (see Theorem 11.7 of Chapter 1):

$$\varphi_k \in H^{p_k}_{00}(\mathbf{R}^n_+) = [H^r_0(\mathbf{R}^n_+), H^0(\mathbf{R}^n_+)]_{(k+1/2)/s} \quad (r > \tfrac{1}{2})$$

(replace $H^r_0(\mathbf{R}^n_+)$ by $H^r_{00}(\mathbf{R}^n_+)$ if $r = \text{integer} + \frac{1}{2}$) and consequently, according to Theorem 4.2, Chapter 1, there exists a $w \in L^2(0, \infty; H^r_0(\mathbf{R}^n_+)) \cap \cap H^s(0, \infty; L^2(\mathbf{R}^n_+))$ such that

$$w^{(k)}(0) = \varphi_k;$$

we can choose the desired function w in this way. □

Remark 2.4. In the sequel, we shall constantly use the spaces

$$H^{2rm,r}(Q), \quad r \geqq 0, \quad \text{integer } m \geqq 1.$$

Then, for $r > \frac{1}{2}$, Theorem 2.1 yields:

$$\text{(2.55)} \qquad \frac{\partial^j u}{\partial \nu^j} \in H^{2rm-j-1/2,(2rm-j-1/2)/2m}(\Sigma), \quad j < 2rm - \frac{1}{2}$$

$$\text{(2.56)} \qquad \frac{\partial^k u}{\partial t^k}(x, 0) \in H^{2m(r-k-1/2)}(\Omega), \quad k < r - \frac{1}{2}.$$

3. Evolution Equations and the Laplace Transform

3.1 Vector Distribution Solutions

Let H be a *Banach* space (later on we shall assume that H is a Hilbert space) and let A be a closed unbounded operator in H such that

$$\text{(3.1)} \qquad \begin{cases} A + p \textit{ is an isomorphism of } D(A) \text{ (norm of the graph)} \to H, \\ \textit{for } p = \xi + i\eta,\ \xi > \xi_0,\ \eta \in \mathbf{R}, \textit{ such that} \\ \|(A + p)^{-1}\|_{\mathscr{L}(H;H)} \leqq \text{polynomial in } |p|. \quad \square \end{cases}$$

The space $\mathscr{D}'_+(H)$

In Chapter 1, Section 1.3, we introduced the space $\mathscr{D}'(H)$ of distributions on $\mathbf{R}_t$ taking their values in H (let $a = -\infty$, $b = +\infty$ in (1.17), Chapter 1), where H is a Hilbert space or a Banach space.

We denote by $\mathscr{D}'_+(H)$ the subspace of $\mathscr{D}'(H)$ of distributions f which vanish for $t < t_f$ (where t_f depends on f), that is such that

$$\langle f, \varphi \rangle = 0 \qquad \forall \varphi \in \mathscr{D} = \mathscr{D}(\mathbf{R}_t),$$

where φ has compact support on $t < t_f$.

It can be shown (see L. Schwartz [6]) that $\mathscr{D}'_+(H) = \mathscr{L}(\mathscr{D}_-; H)$ = space of continuous linear mappings of $\mathscr{D}_-$ into H, where $\mathscr{D}_-$ = space of functions φ of class C^∞ on $\mathbf{R}_t$, with support restricted to the right (i.e. $\varphi(t) = 0$ for $t > t_\varphi$, where t_φ depends on φ). The space $\mathscr{D}_-$ is provided with the inductive limit topology of $\mathscr{D}_-^{(t_n)} = \{\varphi \mid \varphi \in \mathscr{D}_-, \varphi(t) = 0$ for $t > t_n$, t_n fixed$\}$ (the space $\mathscr{D}_-^{(t_n)}$ is provided with the topology of compact convergence of the functions and of each of their derivatives). In general,

$$\mathscr{D}'_+(H) = \mathscr{L}(\mathscr{D}_-; H)$$

is provided with the topology of uniform convergence on the bounded sets of $\mathscr{D}_-$. $\square$

We may now state[1]

Theorem 3.1. *Let* A *be given with* (2.1). *Then, for* f *given in* $\mathscr{D}'_+(H)$, *there exists a unique* $u \in \mathscr{D}'_+(D(A))$ *such that*

$$(3.2) \qquad A\,u + u' = f \quad \text{in} \quad \mathscr{D}'_+(H) \qquad \left(u' = \frac{du}{dt}\right).$$

[Furthermore, $f \to u$ is a continuous linear mapping of $\mathscr{D}'_+(H) \to \to \mathscr{D}'_+(D(A))$].

Proof. 1) We shall show the existence of a distribution $\mathscr{G}$ with the following properties:

$$(3.3) \qquad \mathscr{G} \in \mathscr{D}'_+(\mathscr{L}(H; D(A))),$$

$$(3.4)_1 \qquad \left(A + \frac{\partial}{\partial t}\right) \underset{(t)}{*} \mathscr{G} = \delta \otimes I_H, \quad I_H = \text{identity in } H,$$

$$(3.4)_2 \qquad \mathscr{G} \underset{(t)}{*} \left(A + \frac{\partial}{\partial t}\right) = \delta \otimes I_{D(A)}, \quad I_{D(A)} = \text{identity in } D(A)$$

(where $\underset{(t)}{*}$ denotes *convolution in* t; for convolution of vector-valued distributions we refer the reader to L. Schwartz [6]; we shall only need the elements of the theory here); furthermore

$$(3.5) \qquad \begin{cases} \mathscr{G} \text{ admits a Laplace transform in } t \text{ (in the sense of distributions;} \\ \text{see L. Schwartz [2]) and } \mathscr{G} \text{ has support in } t \text{ on } t \geqq 0. \end{cases}$$

The theorem follows immediately from these properties: the unique solution to the problem is

$$(3.6) \qquad u = \mathscr{G} \underset{(t)}{*} f.$$

2) It remains therefore to construct $\mathscr{G}$ with properties (3.3) to (3.5). Let $\mathscr{G}(p)$ be the Laplace transform of $\mathscr{G}$ ($\mathscr{F}_t$ = Fourier transform in t):

$$(3.7) \qquad \hat{\mathscr{G}}(p) = \mathscr{F}_t(e^{-\xi t}\,\mathscr{G}) = \int_{+\infty}^{-\infty} e^{-i\eta t}(e^{-\xi t}\,\mathscr{G}(t))\,dt \quad \text{(formally)}.$$

Then – assuming the existence of $\mathscr{G}(p)$ for $\xi > \xi_1$ with a suitable ξ_1 – equations $(3.4)_1$ and $(3.4)_2$ become

$$(A + p)\,\hat{\mathscr{G}}(p) = I_H$$

$$\hat{\mathscr{G}}(p)\,(A + p) = I_{D(A)}.$$

---

[1] We shall obtain more general results in Chapter 9 of Volume III.

Therefore, we may take

(3.8) $\hat{\mathscr{G}}(p) = (A + p)^{-1}$, $\xi > \xi_0$ (which exists according to (3.11)),

if the inverse Laplace transform of $(A + p)^{-1}$ exists and has properties (3.3) and (3.5).

But existence and properties (3.3) and (3.5) follow from (3.1), by a result on supports in Laplace transformation (Lions [1]), and from the property:

$p \to (A + p)^{-1}$ is a *holomorphic* mapping of $\xi > \xi_0 \to \mathscr{L}(H; D(A))$.

Indeed, this property is a consequence of the *resolvent identity*: if p and p' satisfy $\operatorname{Re} p > \xi_0$, $\operatorname{Re} p' > \xi_0$, then

$$(p' - p)^{-1}[(A + p')^{-1} - (A + p)^{-1}] = -(A + p')^{-1}(A + p)^{-1},$$

which shows that $\frac{d}{dp}(A + p)^{-1}$ exists (in $\mathscr{L}(H; D(A))$) and is equal to $-(A + p)^{-2}$. Whence the theorem. □

3.2 L^2-Solutions

Theorem 3.2. *Let A be given with* (3.1) *and furthermore satisfy*

(3.9) $$\|(A + p)^{-1}\|_{\mathscr{L}(H;H)} \leqq \frac{c}{|p| + 1} \quad \textit{for} \quad \xi > \xi_0, \qquad c = \text{constant}.$$

Assume that H is a Hilbert space.

Then, for f given in $L^2(0, T; H)$, there exists a unique function u satisfying

(3.10) $$u \in L^2(0, T; D(A)),$$

(3.11) $$A u + u' = f,$$

(3.12) $$u(0) = u_0, \quad u_0 \textit{ given in } [D(A), H]_{1/2}$$

(notation of Chapter 1, Section 2; $u(0)$ is well-defined since $u' \in L^2(0, T; H)$, thanks to (3.11)).

Proof. 1) We can reduce the problem to the case "$u_0 = 0$" (which is *not essential*); indeed, since $u_0 \in [D(A), H]_{1/2}$, there exists (see Chapter 1, Section 3) a w satisfying

(3.13) $$\begin{cases} w \in L^2(0, T; D(A)), \\ w' \in L^2(0, T; H), \\ w(0) = u_0 \end{cases}$$

(and which depends continuously on u_0). Then

$$u - w = \Phi$$

must satisfy

$$A\,\Phi + \Phi' = f - (A\,w + w') \in L^2(0, T; H),$$

$$\Phi \in L^2(0, T; D(A)),$$

$$\Phi(0) = 0.$$

2) Let u be the solution of (3.10) to (3.12) with $f = 0$, $u_0 = 0$. Let $\tilde{u}$ = extension by 0 of u outside $]0, T[$. Then

$$A\,\tilde{u} + (\tilde{u})' = u(T) \otimes \delta(t - T).$$

and according to Theorem 3.1,

$$\tilde{u} = \mathscr{G} \underset{(t)}{*} (u(T) \otimes \delta(t - T)) = \mathscr{G}(t - T)\, u(T).$$

Since $\mathscr{G}(t) = 0$ for $t < 0$, we see that $\mathscr{G}(t - T) = 0$ for $t < T$ and therefore

$$u = \text{restriction of } \tilde{u} \text{ to }]0, T[\; = 0;$$

this shows the *uniqueness*.

3) Now we consider

$$\text{(3.14)} \qquad \begin{cases} U = \mathscr{G} \underset{(t)}{*} \tilde{f} \\ \tilde{f} = \text{extension by 0 of } f \text{ outside }]0,T[. \end{cases}$$

We shall verify that

$$\text{(3.15)} \qquad e^{-\xi t}\, U \in L^2(\mathbf{R}_t; D(A)) \quad \forall \xi > \xi_0,$$

$$\text{(3.16)} \qquad A\,U + U' = \tilde{f} \quad \text{on } \mathbf{R}.$$

Then

$$U' = f - A\,U \in L^2(\mathbf{R}_t; H)$$

and since $U = 0$ for $t < 0$, we shall have

$$\text{(3.17)} \qquad U(0) = 0.$$

Then the *restriction* u of U to $]0, T[$ is a solution to the problem.

Since (3.16) follows from (3.14) and Theorem 3.1, it just remains to verify (3.15). But

$$e^{-\xi t} U = (e^{-\xi t} \mathscr{G}) \underset{(t)}{*} (e^{-\xi t} \tilde{f}),$$

therefore by Fourier transform in t:

$$\mathscr{F}_t(e^{-\xi t} U) = \hat{\mathscr{G}}(p) \mathscr{F}_t(e^{-\xi t} \tilde{f}).$$

Applying (3.9), which is equivalent to

$$\| (A + p)^{-1} \|_{\mathscr{L}(H; D(A))} \leqq c_1, \tag{3.18}$$

we deduce from it, thanks to (3.8):

$$\| \mathscr{F}_t(e^{-\xi t} U)(\eta) \|^2_{D(A)} \leqq c_1^2 \| \mathscr{F}_t(e^{-\xi t} \tilde{f})(\eta) \|^2_H .$$

Integrating in η and applying Plancherel's theorem (*which assumes H to be a Hilbert space*) it follows that, for $\xi > \xi_0$:

$$\int_R \| e^{-\xi t} U(t) \|^2_{D(A)} dt \leqq c_1^2 \int_R \| e^{-\xi t} \tilde{f}(t) \|^2_H dt < \infty;$$

whence (3.15). □

4. The Case of Operators Independent of t

4.1 Hypotheses

We shall again consider the operator P given in (1.1) and (1.2), and assume that the coefficients a_{pq} do not depend on t; thus $a_{pq} = a_{pq}(x) \in \mathscr{D}(\bar{\Omega})$.

We introduce the auxiliary variable $y \in \mathbf{R}_y$ and the operators

$$\Lambda_\theta = A(x, D_x) + e^{i\theta}(-1)^m D_y^{2m}, \quad \theta \in \left[-\frac{\pi}{2}, \frac{\pi}{2}\right]. \tag{4.1}$$

The formal adjoint of Λ_θ in $\Omega \times \mathbf{R}_y$ is

$$\Lambda_\theta^* = A^*(x, D_x) + e^{-i\theta}(-1)^m D_y^{2m}. \tag{4.2}$$

We again consider the boundary operators $\{B_j\}_{j=0}^{m-1}$ given by (1.4) and assume that the coefficients b_{jh} do not depend on t; thus $b_{jh} = b_{jh}(x) \in \mathscr{D}(\Gamma)$.

Finally, we assume that the system $\{\Lambda_\theta, B_j\}$, for all $\theta \in \left[-\frac{\pi}{2}, \frac{\pi}{2}\right]$, is a "regular elliptic" system in $\Omega \times \mathbf{R}_y$; more precisely, we assume that

(i) *Λ_θ is properly elliptic in $\bar{\Omega} \times \mathbf{R}_y$* $\forall \theta \in \left[-\frac{\pi}{2}, \frac{\pi}{2}\right]$,

(ii) *$\{B_j\}_{j=0}^{m-1}$ is normal on Γ and of order m_j, with $0 \leqq m_j \leqq 2m-1$,*

(iii) *$\{B_j\}_{j=0}^{m-1}$ covers Λ_θ on $\Gamma \times \mathbf{R}_y$* $\forall \theta \in \left[-\frac{\pi}{2}, \frac{\pi}{2}\right]$.

Then, applying the elliptic theory of Chapter 2, it follows that the formal adjoint system $\{\Lambda_\theta^*, C_j\}$ is also a "regular elliptic" system (see Chapter 2, Section 2) and that we have the *following estimates*:

(4.3)
$$\begin{cases} \text{there exists a constant } c, \text{ independent of } \theta \in \left[-\dfrac{\pi}{2}, \dfrac{\pi}{2}\right], \text{ such} \\ \text{that } \forall \omega \in H^{2m}(\Omega \times \mathbf{R}_y) \text{ with support in } y \text{ on }]-1, 1[\text{ and such} \\ \text{that } B_j \omega = 0, 0 \leqq j \leqq m-1 \text{ (respectively } C_j \omega = 0, 0 \leqq j \leqq \\ \leqq m-1) \text{ on } \Gamma \times \mathbf{R}_y, \text{ we have} \\ \qquad |||\Lambda_\theta \omega|||_0 + |||\omega|||_0 \geqq c\, |||\omega|||_{2m} \\ \text{(respectively } |||\Lambda_\theta^* \omega|||_0 + |||\omega|||_0 \geqq c\, |||\omega|||_{2m}), \end{cases}$$

where we have used the notation

$$|||\omega|||_k = \text{norm of } \omega \text{ in } H^k(\Omega \times \mathbf{R}_y).$$

(See Chapter 2, Sections 4 and 5; note that, by continuity, we easily see that the constant c is *independent of the parameter* $\theta \in \left[-\frac{\pi}{2}, \frac{\pi}{2}\right]$; see also Chapter 2, Remark 4.1.) □

4.2 Basic Inequalities

We define A (resp. A^*) to be an unbounded operator in $H = L^2(\Omega)$ with

(4.4)
$$D(A) = \left\{v \mid v \in H^{2m}(\Omega), B_j\left(x, \frac{\partial}{\partial x}\right) v = 0 \text{ on } \Gamma, \ 0 \leqq j \leqq m-1\right\}$$

and

(4.5)
$$D(A^*) = \left\{v \mid v \in H^{2m}(\Omega), C_j\left(x, \frac{\partial}{\partial x}\right) v = 0 \text{ on } \Gamma, \ 0 \leqq j \leqq m-1\right\}.$$

Theorem 4.1. *Under the hypotheses of Section* 4.1, *there exists a* $\xi_0 \in \mathbf{R}$ *and an* $\alpha > 0$ *such that,* $\forall p = \xi + i\eta$, $\xi > \xi_0$, *we have*

(4.6) $$\|(A + p)\, v\|_0 \geqq \alpha \|v\|_{2m} \qquad \forall v \in D(A),$$

(4.7) $$\|(A^* + p)\, v\|_0 \geqq \alpha \|v\|_{2m} \qquad \forall v \in D(A^*)$$

(where $\|\ \|_k$ = norm in $H^k(\Omega)$).

Proof. We apply a method of Agmon and Nirenberg.

We consider $v \in D(A)$ (the proof of (4.7) is analogous to the one which follows) and we introduce

$$(4.8)\qquad \begin{cases} w(x, y) = z(y)\, e^{i\mu y} v(x), & \mu \in \mathbf{R}, \\ z \in \mathscr{D}(\mathbf{R}) \text{ with support in }]-1, 1[. \end{cases}$$

Since $v \in D(A)$, the function w satisfies the conditions of (4.3).

Denote by $\| w \|_{2m,0}$ (resp. $\| w \|_{0,2m}$) the norm in $L^2(\mathbf{R}_y; H^{2m}(\Omega))$ (resp. $H^{2m}(\mathbf{R}_y; L^2(\Omega))$); since in $H^{2m}(\Omega \times \mathbf{R}_y)$ we have:

$$|\!|\!| w |\!|\!|_{2m} \sim \| w \|_{2m,0} + \| w \|_{0,2m} \text{ ((1))},$$

the inequality for Λ_θ in (4.3) is equivalent to

$$(4.9)\qquad |\!|\!| \Lambda_\theta w |\!|\!|_0 + |\!|\!| w |\!|\!|_0 \geqq c_1 (\| w \|_{2m,0} + \| w \|_{0,2m}).$$

But

$$\Lambda_\theta w = z(y)\, e^{i\mu y} (A v + (-1)^m e^{i\theta} (i\mu)^{2m} v) +$$

$$+ v \left(\sum_{k=0}^{2m-1} \binom{2m}{k} (i\mu)^k z^{(2m-k)}(y) \right) e^{i\mu y},$$

from which, since $|e^{i\mu y}| = 1$:

$$|\!|\!| \Lambda_\theta w |\!|\!|_0 \leqq c_2 \| (A + e^{i\theta} \mu^{2m}) v \|_0 + c_3 (1 + |\mu|^{2m-1}) \| v \|_0$$

and since

$$|\!|\!| w |\!|\!|_0 \leqq c_2 \| v \|_0$$

we see that (4.9) implies

(4.10)

$$\| (A + e^{i\theta} \mu^{2m}) v \|_0 \geqq c_3 (\| w \|_{2m,0} + \| w \|_{0,2m}) - c_4 (1 + |\mu|^{2m-1}) \| v \|_0,$$

where the constants c_i, here as well as in the sequel, are independent of μ, θ, v.

But

$$\| w \|_{2m,0} = c_2 \| v \|_{2m}$$

and

$$\| w \|_{0,2m} \sim |\!|\!| w |\!|\!|_0 + |\!|\!| D_y^{2m} w |\!|\!|_0;$$

from which it follows that

$$\| w \|_{0,2m} \geqq c_5 |\mu|^{2m} \| v_0 \|_0 - c_6 (1 + |\mu|^{2m-1}) \| v \|_0.$$

((1)) The symbol $\sim$ denotes *equivalent norms*.

Introducing this inequality in (4.10) we obtain:

$$\| (A + e^{i\theta} \mu^{2m}) v \|_0 \geqq c_7 (\| v \|_{2m} + |\mu|^{2m} \| v \|_0) - \\ - c_8 (1 + |\mu|^{2m-1}) \| v \|_0$$

and therefore

$$\| (A + e^{i\theta} \mu^{2m}) v \|_0 \geqq c_9 (\| v \|_{2m} + |\mu|^{2m} \| v \|_0) \tag{4.11}$$

for $\mu \geqq \xi_0$ sufficiently large.

But since this is valid for $\forall \theta \in \left[-\frac{\pi}{2}, \frac{\pi}{2} \right]$, we can choose μ and θ, if

$$p = \xi + i \eta, \quad \xi > \xi_0,$$

so that $e^{i\theta} \mu^{2m} = p$ and therefore

$$\| (A + p) v \|_0 \geqq c_9 (\| v \|_{2m} + |p| \| v \|_0), \tag{4.12}$$

from which (4.6) follows as a particular case.

The proof of (4.7) follows the same lines. □

4.3 Solution of the Problem

Theorem 4.2. *Under the hypotheses of Section* 4.1, *there exists a unique function* $u \in H^{2m,1}(Q)$ *satisfying*

$$\begin{cases} A u + u' = f, & f \text{ given in } L^2(Q) \\ B_j u = 0 \quad \text{on} \quad \Sigma = \Gamma \times]0, T[, & 0 \leqq j \leqq m-1, \\ u(x, 0) = 0. \end{cases} \tag{4.13}$$

Proof. Apply Theorem 3.2 with $D(A)$ given by (4.4) (and $u_0 = 0$). It is sufficient to show that (3.9) holds.

But according to Theorem 4.1 and the *elliptic theory* (Chapter 2, Section 8) it follows from (4.6) and (4.7) that A^*, defined with $D(A^*)$ given by (4.5), is the adjoint of A in the sense of unbounded operators in H and that $(A + p)$ *is an isomorphism of* $D(A) \to H$ for $\xi > \xi_0$.

Furthermore, according to (4.6):

$$\| (A + p)^{-1} \|_{\mathscr{L}(H; D(A))} \leqq \text{constant}.$$

(which is equivalent to (3.9)). □

Remark 4.1. Non-homogeneous case. Let the problem be

$$\begin{cases} A u + u' = f, & f \in L^2(Q), \\ B_j u = g_j \quad \text{on } \Sigma, & 0 \leqq j \leqq m-1, \\ g_j \quad \text{given in} \quad H^{2m-m_j-1/2, (2m-m_j-1/2)/2m}(\Sigma) \\ u(x, 0) = u_0, & u_0 \text{ given in } H^m(\Omega), \end{cases} \tag{4.14}$$

where g_j, u_0 satisfy the *compatibility relations* $(\mathcal{R}.\mathcal{C}.)$:

$$(4.15)\qquad \begin{cases} g_j(x,0) = B_j\left(x, \dfrac{\partial}{\partial x}\right) u_0 \quad \text{on } \Gamma \\ \text{for all } j \text{ such that } m_j \leqq m-1. \end{cases}$$

Thanks to the $(\mathcal{R}.\mathcal{C}.)$ (4.15), there exists a $w \in H^{2m,1}(Q)$ with

$$(4.16)\qquad \begin{cases} B_j w = g_j, \quad 0 \leqq j \leqq m-1, \\ w(x,0) = u_0(x) \end{cases}$$

(in fact the compatibility relations are precisely the necessary and sufficient conditions for the existence of $w \in H^{2m,1}(Q)$, with (4.16); see Section 2.5).

Then $(u - w)$ must satisfy

$$A(u-w) + (u-w)' = f - (Aw + w') \in L^2(Q),$$

$$B_j(u-w) = 0,$$

$$u(0) - w(0) = 0.$$

Therefore the problem is reduced to (4.13) and we have

Theorem 4.3. *Under the hypotheses of Section* 4.1, *if* g_j *and* u_0 *are given with the* $(\mathcal{R}.\mathcal{C}.)$ (4.15), *then problem* (4.14) *admits a unique solution in* $H^{2m,1}(Q)$.

5. Regularity

5.1 Preliminaries

In Section 4, we started with unbounded operators in $H = L^2(\Omega)$.

Now, we may consider (having assumed that the coefficients of A and B_j are sufficiently regular, see Section 4.1) A to be an unbounded operator in $H^k(\Omega)$, arbitrary $k \geqq 0$, *integer or not*. However, we shall assume k to be an integer in order to simplify some technical points.

Thus, we take

$$(5.1)\qquad H = H^k(\Omega),$$

$$(5.2)\qquad D(A) = \{v \mid v \in H^{k+2m}(\Omega), B_j v = 0, 0 \leqq j \leqq m-1\},$$

$$(5.3)\qquad D(A^*) = \{v \mid v \in H^{k+2m}(\Omega), C_j v = 0, 0 \leqq j \leqq m-1\}.$$

In order to arrive at the inequalities which will replace (4.6) and (4.7), we shall use the fact that, under the hypotheses of Section 4.1, we also have (see Chapter 2, Section 4.5) the estimates:

(5.4) $$\begin{cases} \text{for fixed integer } k, \text{ there exists a constant } c, \text{ independent of } \theta, \\ \text{such that for all } w \in H^{k+2m}(\Omega \times \mathbf{R}_y) \text{ with support in } y \text{ on} \\ ]-1, 1[, \text{ and with } B_j w = 0,\ 0 \leqq j \leqq m-1 \text{ (resp. } C_j w = 0, \\ 0 \leqq j \leqq m-1), \text{ on } \Gamma \times \mathbf{R}_y, \text{ we have:} \\ |||\Lambda_\theta w|||_k + |||w|||_k \geqq c\, |||w|||_{k+2m} \\ (\text{resp. } |||\Lambda_\theta^* w|||_k + |||w|||_k \geqq c\, |||w|||_{k+2m}). \\ (\text{in the same notation as in (4.3)}). \end{cases}$$

5.2 Basic Inequalities

Theorem 5.1. *Under the hypotheses of Section 4.1, there exists a $\xi_0 \in \mathbf{R}$ and an $\alpha > 0$ such that $\forall p = \xi + \mathrm{i}\eta$, $\xi > \xi_0$, we have*

(5.5) $$\|(A + p)v\|_k + (1 + |p|^{k/2m})\,\|(A + p)v\|_0 \geqq$$
$$\geqq \alpha(\|v\|_{k+2m} + |p|^{1+k/2m}\,\|v\|_0) \quad \forall v \in D(A)$$

(resp.

(5.6) $$\|(A^* + p)v\|_k + (1 + |p|^{k/2m})\,\|(A^* + p)v\|_0 \geqq$$
$$\geqq \alpha(\|v\|_{k+2m} + |p|^{1+k/2m}\,\|v\|_0) \quad \forall v \in D(A^*)).$$

Proof. The principle of the proof is analogous to the one of Theorem 4.1. Let $v \in D(A)$ (defined by (5.2)). Introduce w by (4.8). In the notation of the proof of Theorem 4.1, (5.4) may be written:

(5.7) $$\begin{cases} \|\Lambda_\theta w\|_{0,k} + \|\Lambda_\theta w\|_{k,0} + \|w\|_{k,0} + \|w\|_{0,k} \geqq \\ \qquad \geqq c_1(\|w\|_{k+2m,0} + \|w\|_{0,k+2m}). \end{cases}$$

But using the explicit form of $\Lambda_\theta w$, we have:

$$\|\Lambda_\theta w\|_{k,0} \leqq c_2\,\|(A + e^{i\theta}\mu^{2m})v\|_k + c_2(1 + |\mu|^{2m-1})\,\|v\|_k$$

and

$$\|\Lambda_\theta w\|_{0,k} \leqq c_3(1 + |\mu|^k)\,\|(A + e^{i\theta}\mu^{2m})v\|_0 + c_3(1 + |\mu|^{k+2m-1})\,\|v\|_0.$$

On the other hand:

$$\|w\|_{k+2m,0} = c_4\,\|v\|_{k+2m}$$

and

$$\|w\|_{0,k+2m} \geqq c_5\,|\mu|^{k+2m}\,\|v\|_0 - c_6(1 + |\mu|^{k+2m-1})\,\|v\|_0$$

and therefore (5.7) yields

$$(5.8)\quad \| (A + e^{i\theta} \mu^{2m}) v \|_k + (1 + |\mu|^k) \| (A + e^{i\theta} \mu^{2m}) v \|_0 \geqq$$
$$\geqq c_7 [\| v \|_{k+2m} + |\mu|^{k+2m} \| v \|_0 -$$
$$- (1 + |\mu|^{k+2m-1}) \| v \|_0 - (1 + |\mu|^{2m-1}) \| v \|_k].$$

But (see Chapter 1, Remark 9.1):

$$\| v \|_k \leqq c_8 \| v \|_{k+2m}^{k/(k+2m)} \| v \|_0^{1-k/(k+2m)},$$

therefore

$$|\mu|^{2m-1} \| v \|_k \leqq \tfrac{1}{2} c_7 \| v \|_{k+2m} + c_9 |\mu|^{(2m-1)(k+2m)/2m} \| v \|_0$$

and consequently (5.8) yields:

$$(5.9)\quad \begin{cases} \| (A + e^{i\theta} \mu^{2m}) v \|_k + (1 + |\mu|^k) \| (A + e^{i\theta} \mu^{2m}) v \|_0 \geqq \\ \geqq c_{10} [\| v \|_{k+2m} + |\mu|^{k+2m} \| v \|_0] \ \text{ for } \mu \geqq \mu_0 \text{ sufficiently large.} \end{cases}$$

Letting, as in Theorem 2.1, $p = e^{i\theta} \mu^{2m}$, (3.5) follows. □

5.3 An Abstract Result

In order to clarify the use of the "basic inequalities", it may be useful to first state an abstract result – a very simple variant of Theorem 3.2.

Let H be a Hilbert space contained in another Hilbert space $\mathscr{H}$:

$$(5.10)\qquad H \subset \mathscr{H} \quad \text{with continuous injection.}$$

Let the given operator A be unbounded in H. We assume:

$$(5.11)\quad \begin{cases} \text{(2.2) holds and furthermore, } \forall u \in D(A), \text{ we have:} \\ \| (A + p) u \|_H + (1 + |p|^\beta) \| (A + p) u \|_{\mathscr{H}} \geqq \\ \qquad \geqq c (\| u \|_{D(A)} + |p|^{1+\beta} \| u \|_{\mathscr{H}}) \\ \forall p \text{ with } \xi > \xi_0, \text{ where } \beta > 0 \text{ is given.} \end{cases}$$

Theorem 5.2. *Let A be given and satisfy* (5.11). *Let f be given with*

$$(5.12)\quad \begin{cases} f \in L^2(0, T; H) \cap H^\beta(0, T; \mathscr{H}), \quad f^{(j)}(0) = 0 \\ \text{if } j < \beta - \tfrac{1}{2}, \quad \beta \neq \text{integer} + \tfrac{1}{2}; \\ \dfrac{1}{\sqrt{t}} f^{(\mu)}(t) \in L^2(0, T; \mathscr{H}) \ \text{ if } \beta = \mu + \dfrac{1}{2}, \quad \mu = \text{integer.} \end{cases}$$

Then there exists a unique u, with

$$u \in L^2(0, T; D(A)) \cap H^{\beta+1}(0, T; \mathscr{H}), \tag{5.13}$$

$$A u + u' = f, \tag{5.14}$$

$$u(0) = 0. \tag{5.15}$$

Proof. Since (3.2) holds, uniqueness follows as in Theorem 4.1.

We *extend* f to $\tilde{f}$ on $\mathbf{R}$ by 0 for $t < 0$ in such a way that $\tilde{f}$ has compact support in t and satisfies

$$\tilde{f} \in H^\beta(\mathbf{R}; \mathscr{H}) \cap L^2(\mathbf{R}; H). \tag{5.16}$$

(Extension "by reflection" about $t = T$; see Chapter 1, Section 2.) As in Theorem 3.1, we solve

$$A U + \frac{dU}{dt} = \tilde{f} \quad \text{on} \quad \mathbf{R}. \tag{5.17}$$

By Laplace transform followed by inversion, we have

$$e^{-\xi t} U = \mathscr{F}_\eta^{-1}[(A + p)^{-1} \mathscr{F}_t(e^{-\xi t} \tilde{f})] \tag{5.18}$$

for $\xi > \xi_0$.

But since $\tilde{f}$ has compact support and satisfies (5.16), we have:

$$e^{-\xi r} \tilde{f} \in H^\beta(\mathbf{R}; \mathscr{H}) \cap L^2(\mathbf{R}; H). \tag{5.19}$$

Let

$$F(p) = \mathscr{F}_t(e^{-\xi t} \tilde{f}).$$

Then, according to (5.11):

$$\| (A + p)^{-1} F(p) \|_{D(A)} \leqq c \| F(p) \|_H + c(1 + |p|^\beta) \| F(p) \|_{\mathscr{H}} \tag{5.20}$$

and according to (5.19):

$$(1 + |\eta|^\beta) F(\xi + i\eta) \in L^2(\mathbf{R}_\eta; \mathscr{H}) \quad \forall \xi. \tag{5.21}$$

Therefore (5.20) shows that

$$\eta \to (A + p)^{-1} F(p), \quad \xi > \xi_0,$$

belongs to $L^2(\mathbf{R}_\eta; D(A))$ and then (5.18) shows that

$$e^{-\xi t} U \in L^2(\mathbf{R}_t; D(A)).$$

Therefore, if u = restriction of U to $]0, T[$, we have:

$$u \in L^2(0, T; D(A)). \tag{5.22}$$

But according to (5.11), we have *an analogous bound for*

$$|p|^{1+\beta} (A + p)^{-1} F(p),$$

from which we obtain that

$$e^{-\xi t} U \in H^{\beta+1}(\mathbf{R}_t; \mathscr{H}) \tag{5.23}$$

and therefore that

$$u \in H^{\beta+1}(0, T; \mathscr{H}),$$

which, together with (5.22), yields the theorem. □

5.4 Solution of the Boundary Value Problem

Theorem 5.3. *Under the hypotheses of Section 4.1, let f, g_j, u_0 be given with*

$$f \in H^{k,k/2m}(Q), \quad \textit{integer} \quad k > 0, \tag{5.24}$$

$$g_j \in H^{k+2m-m_j-1/2,(k+2m-m_j-1/2)/2m}(\Sigma), \tag{5.25}$$

$$u_0 \in H^{k+m}(\Omega), \tag{5.26}$$

and satisfy the compatibility relations (5.31) below. Then there exists a unique function u satisfying

$$u \in H^{k+2m,k/2m+1}(Q), \tag{5.27}$$

$$A u + u' = f, \tag{5.28}$$

$$B_j u = g_j, \quad 0 \leqq j \leqq m-1, \tag{5.29}$$

$$u(0) = u_0. \tag{5.30}$$

Proof. 1) *The compatibility relations* are the "relations needed" (see Section 2) for the existence of $w \in H^{k+2m,k/2m+1}(Q)$, with

$$\begin{cases} B_j w = g_j, \quad w(0) = u_0 \\ (A w + w')^{(j)}(0) = f^{(j)}(0), \quad 0 \leqq j < \dfrac{k}{2m} - \dfrac{1}{2}. \end{cases} \tag{5.31}$$

Then $u - w = \Phi$ satisfies

$$A \Phi + \Phi' = f - (A w + w') \in H^{k,k/2m}(Q),$$

$$B_j \Phi = 0, \quad 0 \leqq j \leqq m-1,$$

$$\Phi(0) = 0,$$

and

$$(f - (A w + w'))^{(j)}(0) = 0, \quad 0 \leqq j < \frac{k}{2m} - \frac{1}{2}.$$

2) Thus we are led back to Theorem 5.2 with

$$H = H^k(\Omega), \quad \mathscr{H} = H^0(\Omega), \quad \beta = \frac{k}{2m}$$

and (5.11) following from (5.5); whence the theorem. □

6. Case of Time-Dependent Operators. Existence of Solutions in the Spaces $H^{2mr,r}(Q)$, Real $r \geqq 1$

6.1 Hypotheses. Statement of the Result

We now consider the operators:

(6.1) $$A(t) = A(x, t, D_x) = \sum_{|p|, |q| \leqq m} (-1)^{|p|} D_x^p (a_{pq}(x, t) D_x^q),$$

(6.2) $$B_j(t) = B_j(x, t, D_x) = \sum_{|h| \leqq m_j} b_{jh}(x, t) D_x^q.$$

We shall make the following *hypotheses*:

(6.3) $\left\{\begin{array}{l} \textit{for every } t = t_0 \in [0, T], \textit{ system } \{A(t_0), B_j(t_0)\} \textit{ satisfies the} \\ \textit{hypotheses of Section } 4.1; \end{array}\right.$

(therefore we shall have:

$$|||(A(t_0) + e^{i\theta}(-1)^m D_y^{2m}) w|||_0 + |||w|||_0 \geqq c_{t_0} |||w|||_{2m}$$
$$\forall w \in H^{2m}(\Omega \times \mathbf{R}_y), \quad \text{with} \quad B_j(t_0) w = 0$$

and the analogous "adjoint" conditions).

We then make the following regularity hypotheses on the dependence on t (these hypotheses can be weakened, as we shall see in Remark 6.1):

(6.4) $$a_{pq} \in C^1([0, T]; \mathscr{D}(\bar{\Omega})) \qquad \forall p, q,$$

(6.5) $$b_{jh} \in C^{2m-m_j+1}([0, T]; \mathscr{D}(\Gamma)) \qquad \forall j, h.$$

Theorem 6.1. *Assume that* (6.3), (6.4) *and* (6.5) *hold. Let* f, g_j *and* u_0 *be given with*

(6.6) $$f \in L^2(Q),$$

(6.7) $$g_j \in H^{2m-m_j-1/2, (2m-m_j-1/2)/2m}(\Sigma), \quad 0 \leqq j \leqq m-1,$$

(6.8) $$u_0 \in H^m(\Omega),$$

and let the following compatibility relations hold:

(6.9) $$B_j\left(x, 0, \frac{\partial}{\partial x}\right) u_0 \bigg|_{\Gamma} = g_j(x, 0) \quad \textit{if} \quad m_j < m - \frac{1}{2}.$$

Then there exists a unique function u *in* $H^{2m,1}(Q)$ *satisfying*

(6.10) $$A u + u' = f,$$

(6.11) $$B_j u = g_j, \quad 0 \leqq j \leqq m-1,$$

(6.12) $$u(0) = u_0.$$

The proof of this theorem (*which contains Theorem* 1.1) will be given in two steps.

6.2 Local Result in t

Let Δt be an interval to be chosen "sufficiently small" later on. Set

$$Q_{\Delta t} = \Omega \times]0, \Delta t[, \quad \Sigma_{\Delta t} = \Gamma \times]0, \Delta t[.$$

Define

(6.13)

$$K_{\Delta t} = \left\{ L^2(Q_{\Delta t}) \times \prod_{j=0}^{m-1} H^{2m-m_j-1/2,(2m-m_j-1/2)/2m}(\Sigma_{\Delta t}) \times H^m(\Omega), \mathscr{R}.\mathscr{C}. \right\}$$

that is

$$K_{\Delta t} = \Big\{ \{f, g_j, u_0\}, f \in L^2(Q_{\Delta t}), g_j \in H^{2m-m_j-1/2,(2m-m_j-1/2)/2m}(\Sigma_{\Delta t}), \\ u_0 \in H^m(\Omega) \text{ and } (6.9) \text{ is satisfied} \Big\}.$$

Let R and R_0 be the operators

$$R: u \to \{A(t)u + u', B_j(t)u, u(x,0)\}, \tag{6.14}$$

$$R_0: u \to \{A(0)u + u', B_j(0)u, u(x,0)\}, \tag{6.15}$$

which *both* satisfy

$$R, R_0 \in \mathscr{L}(H^{2m,1}(Q_{\Delta t}); K_{\Delta t}). \tag{6.16}$$

In the cylinder $Q_{\Delta t}$ ("local" case in t), problem (6.10), (6.11), (6.12) is equivalent to

$$Ru = \{f, g_j, u_0\} = \chi \text{ given in } K_{\Delta t}. \tag{6.17}$$

But thanks to Theorem 4.3, R_0 is an *isomorphism* of $H^{2m,1}(Q_{\Delta t})$ onto $K_{\Delta t}$ for all Δt, and furthermore, according to the estimates,

$$\|R_0^{-1}\|_{\mathscr{L}(K_{\Delta t}; H^{2m,1}(Q_{\Delta t}))} \leqq (\text{constant independent of } \Delta t) = c_1 \tag{6.18}$$ [(1)]

Problem (6.17) is equivalent to

$$(R_0 + (R - R_0))u = \chi,$$

that is to

$$(1 + R_0^{-1}(R - R_0))u = R_0^{-1}\chi. \tag{6.19}$$

From this fact we shall deduce:

Lemma 6.1. *Hypotheses of Theorem* 6.1. *There exists a* Δt *such that problem* (6.10), (6.11), (6.12) *has a unique solution in* $\Omega \times]0, \Delta t[$. *Further-*

[(1)] Note that the bounds in the case of time-independent coefficients hold with constants ***independent*** of the length of the interval in t.

more Δt does not depend on the origin of t, in other words, problem

$$A(t)\,w + w' = f \quad in \quad \Omega \times]t_0, t_0 + \Delta t[,$$
$$B_j w = g_j \quad on \quad \Gamma \times]t_0, t_0 + \Delta t[,$$
$$w(x, t_0) = u_{t_0}(x),$$

with

$$B_j\left(x, t_0, \frac{\partial}{\partial x}\right) u_{t_0}(x) = g_j(x, t_0) \text{ if } m_j < m - \frac{1}{2},$$

admits a unique solution for every $t_0 \in [0, T]$.

Proof. We need to find an estimate for

$$\varphi(\Delta t) = \|R_0^{-1}(R - R_0)\|_{\mathscr{L}(H^{2m,1}(Q_{\Delta t}); H^{2m,1}(Q_{\Delta t}))}.$$

According to (6.18), we have:

$$\varphi(\Delta t) \leqq c_1 \|R - R_0\|_{\mathscr{L}(H^{2m,1}(Q_{\Delta t}); K_{\Delta t})} = c_1\,\psi(\Delta t). \tag{6.20}$$

Let

$$B_1 = \{u \mid u \in H^{2m,1}(Q_{\Delta t}),\ \|u\|_{H^{2m,1}(Q_{\Delta t})} \leqq 1\}.$$

Then

$$\psi(\Delta t) = \sup_{u \in B_1} \|R\,u - R_0\,u\|_{K_{\Delta t}}$$
$$= \sup_{u \in B_1}\left[\|(A(t) - A(0))\,u\|_{H^0(Q_{\Delta t})} + \right.$$
$$\left. + \sum_{j=0}^{m-1} \|(B_j(t) - B_j(0))\,u\|_{H^{\mu_j, \mu_j/2m}(\Sigma_{\Delta t})}\right]$$

(where $\mu_j = 2m - m_j - \frac{1}{2}$).

But thanks to (6.4) and (6.5) we have:

$$\|(A(t) - A(0))\,u\|_{H^0(Q_{\Delta t})} \leqq c_2\,\Delta t\,\|u\|_{H^{2m,1}(Q_{\Delta t})} \quad \text{if} \quad t \in [0, \Delta t] \tag{6.21}$$

and

$$\|(B_j(t) - B_j(0))\,u\|_{H^{\mu_j, \mu_j/2m}(\Sigma_{\Delta t})} \leqq c_3\,\Delta t\,\|u\|_{H^{2m,1}(Q_{\Delta t})} \quad \text{if} \quad t \in [0, \Delta t], \quad \forall j. \tag{6.22}$$

Remark 6.1. Hypotheses (6.4) and (6.5) are only *simple sufficient conditions* implying (6.21) and (6.22); they may be weakened by using the spaces of multipliers on the Sobolev spaces. ☐

Remark 6.2. All the constants are independent of the origin in t. ☐

From (6.20), (6.21), (6.22) we deduce:

$$\varphi(\Delta t) \leqq c_4\,\Delta t$$

and therefore

$$\varphi(\Delta t) < 1 \quad \text{for} \quad \Delta t < \frac{1}{c_4}.$$

Then $1 + R_0^{-1}(R - R_0)$ is invertible in $H^{2m,1}(Q_{\Delta t})$ and Lemma 6.1 follows.

6.3 Proof of Theorem 6.1.

The theorem is an immediate consequence of Lemma 6.1. Choose Δt as in the lemma. Solve the problem in $\Omega \times]0, \Delta t[$; let u^1 be the solution. Then consider u^2, solution (which exists according to Lemma 6.1) of

$$A\, u^2 + (u^2)' = f \quad \text{in} \quad \Omega \times]\Delta t, 2\Delta t[,$$
$$B_j\, u^2 = g_j \quad \text{on} \quad \Gamma \times]\Delta t, 2\Delta t[,$$
$$u^2(x, \Delta t) = u^1(x, \Delta t);$$

the compatibility relations on $\Gamma \times \{t = \Delta t\}$ hold since u^1 is the solution of the problem in $\Omega \times]0, \Delta t[$; thus u^2 exists and so on. Taking $u = u^1$ in $\Omega \times]0, \Delta t[$, u^2 in $\Omega \times]\Delta t, 2\Delta t[$, etc. we obtain the *existence* of a solution.

Uniqueness also results from Lemma 6.1; if u satisfies (6.10), (6.11), (6.12) with $f = 0$, $g_j = 0$, $u_0 = 0$, then $u = 0$ in $\Omega \times]0, \Delta t[$ according to Lemma 6.1; then $u(x, \Delta t) = 0$ implies $u = 0$ in $\Omega \times]\Delta t, 2\Delta t[$, etc. □

6.4 Regular Non-Homogeneous Problems

The same method as in Section 6.3, together with Theorem 5.3, yields the following result. Under the hypotheses of Section 1.1, let

(6.23) $\qquad r \geqq 0$ be given with $2rm = $ *integer*;

let g_j, u_0 and f be given with

(6.24) $\qquad g_j \in H^{2(r+1)m - m_j - 1/2, (r+1) - (m_j + 1/2)/2m}(\Sigma), \quad 0 \leqq j \leqq m-1,$

(6.25) $\qquad u_0 \in H^{2(r+1/2)m}(\Omega),$

(6.26) $\qquad f \in H^{2rm, r}(Q),$

and with the *compatibility relation* ($\mathscr{R}.\mathscr{C}.$), which may be expressed by:

(6.27)
$$\begin{cases} \text{there exists a } w \in H^{2(r+1)m, r+1}(Q) \text{ with } B_j w = g_j \\ (B_j = B_j(t) \text{ now depends on } t),\ 0 \leqq j \leqq m-1, \\ w(x, 0) = u_0(x), \quad x \in \Omega, \\ D_t^k [A(x, t, D_x)\, w + D_t w]\,|_{t=0} = D_t^k f(x, 0), \\ 0 \leqq k < r - \frac{1}{2}.^{(1)} \end{cases}$$

Then problem (6.10), (6.11), (6.12) *admits a unique solution in the space* $H^{2(r+1)m, r+1}(Q)$ (solution which depends continuously – in the

(1) Of course, with the help of Theorem 2.3, we can write these relations explicitly.

obvious topologies – on the data $\{g_j, u_0, f\}$). We shall now extend this to the case "non-integer $2rm$" (with a condition on r):

Theorem 6.2. *Let real $r \geqq 0$, with*

$$2rm \quad \text{and} \quad r \neq \text{integer} + \tfrac{1}{2}. \tag{6.28}$$

Assume that the hypotheses of Section 1.1 *are satisfied. Let $\{f, g_j, u_0\}$ be given with* (6.24), (6.25), (6.26) *and satisfy the* ($\mathcal{R}.\mathcal{C}.$) (6.27). *Then problem* (6.10), 6.11), (6.12) *admits a unique solution u in $H^{2(r+1)m,r+1}(Q)$.*

The mapping $\{f, g_j, u_0\} \to u$ is a continuous mapping of

$$H^{2rm,r}(Q) \times \prod_{j=0}^{m-1} H^{2(r+1)m-m_j-1/2,(r+1)-(m_j+1/2)/2m}(\Sigma) \times$$
$$\times H^{2(r+1/2)m}(\Omega) = \mathcal{F}_r \to H^{2(r+1)m,r+1}(Q).$$

Remark 6.3. We may also state:

the mapping $u \to \left\{\dfrac{\partial u}{\partial t} + A\,u, B_j\,u, u(x, 0)\right\}$ is an isomorphism of $H^{2(r+1)m,r+1}(Q)$ *onto the subspace of the product space $\mathcal{F}_r$ of elements $\{f, g_j, u_0\}$ which satisfy the* ($\mathcal{R}.\mathcal{C}.$). □

Proof of Theorem 6.2. Since, by hypotheses, there exists a w with (6.27), it all comes down to solving (setting $\Phi = u - w$):

$$\begin{cases} \dfrac{\partial \Phi}{\partial t} + A\,\Phi = \psi, \\ B_j\,\Phi = 0, \quad 0 \leqq j \leqq m-1, \\ \Phi(x, 0) = 0 \end{cases} \tag{6.29}$$

the function $\psi\left(= f - \left(\dfrac{\partial w}{\partial t} + A\,w\right)\right)$ satisfies

$$\psi \in H^{2rm,r}(Q), \quad \psi_t^{(k)}(x, 0) = 0, \quad 0 \leqq k < r - \tfrac{1}{2}. \tag{6.30}$$

We have to show that $\Phi \in H^{2(r+1)m,r+1}(Q)$.

But this result is known for integer $2rm$; so let us take r_1 with $r_1 > r$, $2r_1 m$ = integer and let us call G^r the space defined by (6.30). Thus we know that the mapping

$$\psi \to \Phi$$

is a continuous linear mapping of $G^0 \to H^{2m,1}(Q)$ and of $G^{r_1} \to \to H^{2(r_1+1)m,r_1+1}(Q)$. Therefore, *by interpolation*, it is a continuous mapping of

$$[G^{r_1}, G^0]_\theta \to [H^{2(r_1+1)m,r_1+1}(Q), H^{2m,1}(Q)]_\theta.$$

According to Proposition 2.1, the second space is equivalent to

$$H^{2((1-\theta)r_1+1)m,(1-\theta)r_1+1}(Q).$$

Choose θ with $(1-\theta)\,r_1 = r$. It remains to see that $[G^{r_1}, G^0]_\theta = G^r$, $2r\,m$ and $r \neq$ integer $+ \frac{1}{2}$. The proof of this is analogous to the proof of Theorem 11.6. of Chapter 1. □

Remark 6.4. We can also deal with the exceptional values of the parameter r by considering the *integral compatibility relations*. If we do not introduce these integral relations, Theorem 6.2 is no longer valid for r exceptional. Indeed, let us take $f = 0$ in the statement of Theorem 6.2, and in order to simplify the notation, let us replace $r+1$ by r. Then if

$$2(r-\tfrac{1}{2})\,m = \text{integer} + \tfrac{1}{2},$$

or if

$$\frac{2r\,m - m_j - \frac{1}{2}}{2m} = \text{integer} + \frac{1}{2},$$

we shall see that the following inequality *does not hold* (see Solonnikov [2]):

$$(6.31)\qquad \|u\|_{H^{2rm,r}(Q)} \leqq c_1\left(\sum_{j=0}^{m-1} \|g_j\|_{H^{2rm-m_j-1/2,(2rm-m_j-1/2)/2m}(\Sigma)} + \right.$$
$$\left. + \|u_0\|_{H^{2(r-1/2)m}(\Omega)}\right)$$

for $g_j \in \mathscr{D}(\Sigma)$ and $u_0 \in \mathscr{D}(\Omega)$.

In order to simplify somewhat, assume $\Sigma = \Gamma \times]0,\infty[$, $Q = \Omega \times]0,\infty[$.

First we consider $u_0 = 0$. Let $\tilde{g}_j$ (resp. $\tilde{u}$) be the extension of g_j (resp. u) to $\Gamma \times \mathbf{R}$ (resp. $\Omega \times \mathbf{R}$) by 0 for $t < 0$. According to the trace theorem of Section 2, we have:

$$\sum_j \|\tilde{g}_j\|_{H^{2rm-m_j-1/2,(2rm-m_j-1/2)/2m}(\Gamma\times\mathbf{R})} \leqq c_2\,\|\tilde{u}\|_{H^{2rm,r}(\Omega\times\mathbf{R})}$$

and since $r \neq$ integer $+\frac{1}{2}$, we have

$$\|\tilde{u}\|_{H^{2rm,r}(\Omega\times\mathbf{R})} \leqq c_3\|u\|_{H^{2rm,r}(Q)}$$

and therefore if the inequality (6.31) would hold, we would have

$$\sum_{j=0}^{m-1} \|\tilde{g}_j\|_{H^{2rm-m_j-1/2,(2rm-m_j-1/2)/2m}(\Gamma\times\mathbf{R})} \leqq$$
$$\leqq c_3 \sum_{j=0}^{m-1} \|g_j\|_{H^{2rm-m_j-1/2,(2rm-m_j-1/2)/2m}(\Sigma)}$$

which is false (see Chapter 1, Section 11.3) when

$$\frac{2r\,m - m_j - \frac{1}{2}}{2m} = \text{integer} + \frac{1}{2}.$$

The same proof applies when $g_j = 0$ and $u_0 \neq 0$, by considering the extension by 0 outside Ω. □

We make the following additional remarks:

(i) the preceding counter-example extends (same type of proof) to the case of L^p, $p \neq 2$;

(ii) the "explanation" for the preceding counter-example is the following: for the exceptional values of the parameter, the natural topology on the space of elements $\{g_j, u_0\}$ satisfying the $(\mathcal{R}.\mathcal{C}.)$ is *strictly finer* than the one induced by

$$\prod_{j=0}^{m-1} H^{2rm-m_j-1/2(2rm-m_j-1/2)/2m}(\Sigma) \times H^{2(r-1/2)m}(\Omega),$$

(See Chapter 1, Section 11 and Section 2 of this chapter.)

Thus if we work in

$$H^{2rm-m_j-1/2,(2rm-m_j-1/2)/2m}_{0,0}(\Sigma)$$

(space obtained by interpolation) or in $H^{2(r-1/2)m}_{00}(\Omega)$, the preceding considerations *yield no counter-example.* □

7. Adjoint Isomorphism of Order r

7.1 The Adjoint Problem

We shall always assume the hypotheses of Section 1.1 *to be satisfied.*

Replacing P by P^* and the boundary conditions pertaining to B_j by the "adjoint" conditions pertaining to C_j, we deduce the following result from Theorem 6.2: let φ be given in $H^{2(r-1)m,r-1}_{0,0}(Q)$, real $r \geqq 1$, $2rm$ and $r \neq$ integer $+ \frac{1}{2}$; let v be the solution of

$$A^* v - v' = \varphi \quad \text{in } Q, \tag{7.1}$$

$$C_j v = 0 \quad \text{on } \Sigma, \quad 0 \leqq j \leqq m-1, \tag{7.2}$$

$$v(x, T) = 0, \quad x \in \Omega. \tag{7.3}$$

Then

$$v \in H^{2rm,r}(Q). \tag{7.4}$$

In fact, as φ describes $H^{2(r-1)m,r-1}_{0,0}(Q)$, v describes a *subspace* of $H^{2rm,r}(Q)$.

Definition 7.1. *If real* $r \geqq 1$ *with* $2rm$ *and* $r \neq$ *integer* $+ \frac{1}{2}$, *we denote by* $X^r(Q)$ *the space described by the solutions* v *of* (7.1), (7.2), (7.3) *as* φ *describes* $H^{2(r-1)m,r-1}_{0,0}(Q)$.

Therefore, by (7.4), we have

$$X^r(Q) \subset H^{2rm,r}(Q). \tag{7.5}$$

Equivalently, we can define:

$$X^r(Q) = \{v \mid v \in H^{2rm,r}(Q),\, C_j v = 0,\, j = 0, \ldots, m-1,\, v(x,T) = 0,$$
$$A^* v - v' \in H^{2(r-1)m,r-1}_{0,0}(Q)\}. \quad \square$$

7.2 Adjoint Isomorphism of Order r

Providing $X^r(Q)$ with the norm of the graph

$$\|v\|_{X^r(Q)} = (\|v\|^2_{H^{2rm,r}(Q)} + \|P^* v\|^2_{H^{2(r-1)m,r-1}(Q)})^{1/2},$$

we have

(7.6) $\begin{cases} \textit{for } r \geqq 1 \textit{ and with } 2rm \textit{ and } r \neq \textit{integer} + \frac{1}{2},\ P^* \left(= A^* - \dfrac{\partial}{\partial t}\right) \\ \textit{is an isomorphism of } X^r(Q) \textit{ onto } H^{2(r-1)m,r-1}_{0,0}(Q). \end{cases}$

This is what we call the "adjoint isomorphism of order r". □

Remark 7.1. The choice of the space $H^{2(r-1)m,r-1}_{0,0}(Q)$ (as "domain" of φ) is, of course, *completely arbitrary*. We have considered this case for the following reasons:

(i) $H^{2(r-1)m,r-1}_{0,0}(Q)$ is a space for which we have a "*natural*" *regularity theorem*;

(ii) $\mathscr{D}(Q)$ is dense in $H^{2(r-1)m,r-1}_{0,0}(Q)$.

In fact, in Chapter 2, Section 6 (see (6.2)), we introduced $X^r(\Omega)$ with $A^* v \in H^r_0(\Omega)$ for analogous reasons; here, the space $H^{2(r-1)m,r-1}_{0,0}(Q)$ "replaces" the space $H^r_0(\Omega)$ of the elliptic theory. □

We shall now transpose the isomorphism (7.6).

8. Transposition of the Adjoint Isomorphism of Order r. (I): Generalities

8.1 Transposition

We still assume that the hypotheses of Section 1.1 are satisfied. By transposition we deduce from (7.6):

(8.1) $\begin{cases} \text{let } v \to L(v) \text{ be a continuous antilinear form on } X^r(Q)\text{; there} \\ \text{exists a unique } u \in (H^{2(r-1)m,r-1}_{0,0}(Q))' \text{ such that } \langle u, \overline{A^* v - v'}\rangle \\ = L(v) \quad \forall v \in X^r(Q), \end{cases}$

where the brackets $\langle\, ,\, \rangle$ denote the duality between

$$(H^{2(r-1)m,r-1}_{0,0}(Q))' \quad \text{and} \quad H^{2(r-1)m,r-1}_{0,0}(Q). \quad \square$$

8.2 Orientation

We shall now – until Section 12 inclusively – specify and interpret (8.1).

For this purpose, we shall:

(i) interpret the space $(H^{2(r-1)m,r-1}_{0,0}(Q))'$; Section 8.3;
(ii) choose L; Sections 8.4, 9 and 11;
(iii) state trace theorems; Section 10;
(iv) interpret (8.1); Section 12. □

8.3 The Spaces $H^{-\alpha,-\beta}(Q)$, $H^{-\alpha,-\beta}(\Sigma)$, $\alpha, \beta \geqq 0$

Definition 8.1. *Let $\alpha, \beta \in \mathbf{R}$ and $\alpha, \beta \geqq 0$. We define:*

$$H^{-\alpha,-\beta}(Q) = \text{dual of } H^{\alpha,\beta}_{0,0}(Q) \tag{8.2}$$

(where we recall that $H^{\alpha,\beta}_{0,0}(Q)$ = closure of $\mathscr{D}(Q)$ in $H^{\alpha,\beta}(Q)$).

If β is an integer, we obtain from (2.1) and the Hahn-Banach theorem that every continuous antilinear form on $H^{\alpha,\beta}_{0,0}(\Omega \times \mathbf{R}_t)$ may be written (non-uniquely) as:

$$(\psi, v) = (\psi_0, v) + (\psi_1, D_t^\beta v), \tag{8.3}$$

where

$$\psi_0 \in L^2(\mathbf{R}_t; H^{-\alpha}(\Omega)) \quad (H^{-\alpha}(\Omega) = \text{dual of } H_0^\alpha(\Omega)),$$

(see Chapter 1, Section 12) and where

$$\psi_1 \in L^2(\mathbf{R}_t; H^0(\Omega)).$$

Then $\psi \in H^{-\alpha,-\beta}(Q)$ may be represented by:

$$\begin{cases} \psi = \psi_0 + D_t^\beta \psi_2, \\ \psi_0 \in L^2(0, T; H^{-\alpha}(\Omega)), \\ \psi_2 \in L^2(0, T; H^0(\Omega)). \quad \square \end{cases} \tag{8.4}$$

If β *is not an integer*, we may still write a formula of type (8.4) by using *fractional derivatives in t*. □

In the same way, we define $H^{-\alpha,-\beta}(\Sigma)$ = dual of $H^{\alpha,\beta}_{,0}(\Sigma)$ and, for integer β (with fractional derivatives for $\beta \neq$ integer), every $\psi \in H^{-\alpha,-\beta}(\Sigma)$ may be represented, non-uniquely, by

$$\begin{cases} \psi = \psi_0 + D^\beta \psi_2, \\ \psi_0 \in L^2(0, T; H^{-\alpha}(\Gamma)), \\ \psi_2 \in L^2(0, T; H^0(\Gamma)) \quad (= H^0(\Sigma)). \end{cases} \tag{8.5}$$

8.4 (Formal) Choice of L

With the notions of Section 8.3, we may state (8.1) in the form:

(8.6) let L be a continuous antilinear form on $X^r(Q)$; there exists a unique

$$u \in H^{-2(r-1)m, -(r-1)}(Q),$$

such that

$$\langle u, \overline{A^* v - v'} \rangle = L(v) \qquad \forall v \in X^r(Q).$$

Furthermore

(8.7) $L \to u$ is a continuous linear mapping of $(X^r(Q))'$ (antidual of $X^r(Q)) \to H^{-2(r-1)m, -(r-1)}(Q)$. □

We now FORMALLY choose (in the sequel we shall be concerned with giving meaning to this!):

$$L(v) = \int_Q f \bar{v}\, dx\, dt + \sum_{j=0}^{m-1} \int_\Sigma g_j \overline{T_j v}\, d\Sigma + \int_\Omega u_0 \overline{v(x, 0)}\, dx, \tag{8.8}$$

where f, g_j and u_0 are "functions" given in Q, on Σ and in Ω.

We shall "show" (this will also be justified in the sequel) that

$$A u + u' = f \quad \text{in} \quad Q, \tag{8.9}$$

$$B_j u = g_j, \quad 0 \leqq j \leqq m - 1, \tag{8.10}$$

$$u(x, 0) = u_0. \tag{8.11}$$

Indeed, we first apply (8.6) and (8.8) with $v \in \mathcal{D}(Q)$. Then the integrals on Σ and Ω vanish and there remains

$$\langle u, \overline{A^* v} - \bar{v}' \rangle = \int_Q f \bar{v}\, dx\, dt \qquad \forall v \in \mathcal{D}(Q)$$

therefore (8.9) in the sense of distributions on Q.

Then, *assuming that Green's formula holds* (which will be shown later on), we deduce from (1.12a) that

$$(P u, v) \left(= \int_Q f \bar{v}\, dx\, dt \right) = (u, P^* v) + (\vec{S} u, \vec{C} v)_\Sigma - (\vec{B} u, \vec{T} v)_\Sigma + $$
$$+ (u(T), v(T))_\Omega - (u(0), v(0))_\Omega;$$

but if $v \in X^r(Q)$, we have:

$$\vec{C} v = 0 \quad \text{and} \quad v(0) = 0,$$

therefore

$$(8.12) \qquad (u, P^* v) = \int_Q f \bar{v}\, dx\, dt + (\vec{B} u, \vec{T} v)_\Sigma + (u(0), v(0))_\Omega$$

and comparing with (8.6)–(8.8), we obtain (8.10) and (8.11). □

Now the problem is to choose f, g_j and u_0, and then to justify the preceding formal considerations.

9. Choice of f. The Spaces $\Xi^{2rm,r}(Q)$

Compare this section with Chapter 2, Section 6.3.

9.1 The Space $\Xi^{2rm,r}(Q)$

Let ϱ be the function introduced in Chapter 2, Section 6.3 (function "equivalent" to the distance $d(x, \Gamma)$ from x to Γ).

We recall (Chapter 2, Section 6.3) the definition of $\Xi^\mu(\Omega)$, integer $\mu \geqq 1$:

$$(9.1) \qquad \Xi^\mu(\Omega) = \{v \mid \varrho^{|\alpha|} D^\alpha v \in L^2(\Omega), |\alpha| \leqq \mu\},$$

which is a Hilbert space with the norm

$$(9.2) \qquad \| v \|_{\Xi^\mu(\Omega)} = \left(\sum_{|\alpha| \leqq \mu} \| \varrho^{|\alpha|} D^\alpha v \|^2_{L^2(\Omega)} \right)^{1/2}.$$

Next, let $d(t)$ be a fixed infinitely differentiable function on $[0, T]$ such that

$$(9.3) \qquad d(t) = \begin{cases} t & \text{if } t \leqq t_0 \\ T - t & \text{if } T - t_0 \leqq t \leqq T, \end{cases}$$

t_0 fixed with $0 < t_0 < T - t_0$.

Let us start with the case "integer r". We define

$$(9.4) \qquad \Xi^{2rm,r}(Q) = \{v \mid d^j(t)\, v^{(j)} \in L^2(0, T; \Xi^{2(r-j)m}(\Omega)), 0 \leqq j \leqq r\}.$$

Provided with 'the norm

$$(9.5) \qquad \| v \|_{\Xi^{2rm,r}(Q)} = \left(\sum_{j=0}^{r} \| d^j v^{(j)} \|^2_{L^2(0,T;\Xi^{2(r-j)m}(\Omega))} \right)^{1/2}.$$

it is a Hilbert space. □

Proposition 9.1. *The space $\mathscr{D}(Q)$ is dense in $\Xi^{2rm,r}(Q)$.*

Proof. 1) It is sufficient to show that the space of elements $v \in \Xi^{2rm,r}(Q)$ and having *compact support* in Q is dense in $\Xi^{2rm,r}(Q)$ (then we regularize).

2) Let $\delta_\varepsilon(x)$ be a sequence of functions belonging to $\mathscr{D}(\Omega)$ and $\varphi_\varepsilon(t)$ a sequence of functions belonging to $\mathscr{D}(]0, T[)$ such that

$$\begin{aligned} \delta_\varepsilon(x) &= 0 \quad \text{if} \quad d(x, \Gamma) \leqq \varepsilon, \\ \delta_\varepsilon(x) &= 1 \quad \text{if} \quad d(x, \Gamma) \geqq 2\varepsilon, \\ \varphi_\varepsilon(t) &= 0 \quad \text{if} \quad t \leqq \varepsilon \quad \text{or} \quad t \geqq T - \varepsilon, \\ \varphi_\varepsilon(t) &= 1 \quad \text{if} \quad t \geqq 2\varepsilon \quad \text{and} \quad t \leqq T - 2\varepsilon, \end{aligned}$$

and such that

$$\text{(9.6)} \quad \begin{cases} |d(x, \Gamma)^{|\alpha|} \, D^\alpha \delta_\varepsilon(x)| \leqq \text{constant}, & |\alpha| \leqq 2rm, \\ |t^\beta \varphi^{(\beta)}(t)| + |(T-t)^\beta \varphi^{(\beta)}(t)| \leqq \text{constant}, & \beta \leqq r. \end{cases}$$

Such sequences exist.

Let $u \in \Xi^{2rm,r}(Q)$; we introduce:

$$\text{(9.7)} \qquad u_\varepsilon = \delta_\varepsilon(x)\, \varphi_\varepsilon(t)\, u.$$

It is sufficient to show that $u_\varepsilon \to u$ in $\Xi^{2rm,r}(Q)$ as $\varepsilon \to 0$, therefore that

$$d(t)^j \frac{d^j u_\varepsilon}{dt^j} \to d(t)^j \frac{d^j u}{dt^j} \quad \text{in} \quad L^2(0, T; \Xi^{2(r-j)m}(Q)),$$

therefore finally that

$$\text{(9.8)} \quad \begin{cases} \varrho(x)^{|\alpha|}\, d(t)^j D_t^j D_x^\alpha u_\varepsilon \to \varrho(x)^{|\alpha|}\, d(t)^j D_t^j D_x^\alpha u \\ \text{in} \quad L^2(\Omega \times]0, T[) \quad \text{as} \quad \varepsilon \to 0, \quad |\alpha| \leqq 2(r-j)\, m. \end{cases}$$

Replacing u_ε with its value (9.7), we see that we shall have (9.8) if

$$\text{(9.9)} \quad \begin{cases} \varrho^{|\alpha|}\, d(t)^j \varphi_\varepsilon^{(k)}\, (D_x^\beta \delta_\varepsilon)\, (D_t^{j-k} D_x^\gamma u) \to 0 \quad \text{in} \quad L^2(Q), \\ |\alpha| = |\beta| + |\gamma| \leqq 2(r-j)\, m, \qquad k + |\beta| \geqq 1. \end{cases}$$

But the functions appearing in (9.9) may be written:

$$(\varrho^{|\beta|} D_x^\beta \delta_\varepsilon)\, (d(t)^k \varphi_\varepsilon^{(k)})\, (d(t)^{j-k} \varrho^{|\gamma|} D_t^{j-k} D_x^\gamma u);$$

according to (9.6), the first two terms are *bounded* and furthermore tend towards zero a.e. as $\varepsilon \to 0$, from which (9.9) follows by an application of the Lebesgue theorem. □

For the case "non-integer r", we *define* $\Xi^{2rm,r}(Q)$ as follows:

$$\text{(9.10)} \quad \begin{cases} \text{if} \quad r = k + 1 - \theta, \quad 0 < \theta < 1, \quad \text{integer} \quad k \geqq 0, \\ \Xi^{2m,r}(Q) = [\Xi^{2(k+1)m,k+1}(Q), \quad \Xi^{2km,k}(Q)]_\theta. \quad \square \end{cases}$$

Proposition 9.1 *is still true for non-integer* r, thanks to the properties of the interpolation spaces $[\ ,\ ]_\theta$ (see Chapter 1, Section 2). $\square$

Proposition 9.2.

$$X^r(Q) \subset \Xi^{2rm,r}(Q) \quad \textit{for} \quad r \geqq 1, \quad 2rm \quad \textit{and} \quad r \neq \textit{integer} + \tfrac{1}{2}.$$

Proof. Follows immediately, since (see Section 7) $X^r(Q) \subset H^{2rm,r}(Q)$ and obviously $H^{2rm,r}(Q) \subset \Xi^{2rm,r}(Q)$. $\square$

Remark 9.1. (Compare with Chapter 1, Section 6).

We have sought "the smallest possible" space which has the two properties of Propositions 9.1 and 9.2 and which furthermore is *independent of the boundary operators* B_j (in order not to increase the technical difficulties). This has led us to the spaces $\Xi^{2rm,r}(Q)$. $\square$

9.2 The Space $\Xi^{-2rm,-r}(Q)$

By *definition*, we set:

$$\Xi^{-2rm,-r}(Q) = (\Xi^{2rm,r}(Q))', \quad r \geqq 0. \tag{9.11}$$

Proposition 9.3. *The space* $\Xi^{-2rm,-r}(Q)$, *for integer* $r \geqq 0$, *coincides with the space of distributions on* Q *which may be written* (non-uniquely):

$$\begin{cases} f = \sum_{\alpha,j} D_x^\alpha D_t^j (\varrho^{|\alpha|}(x)\, d^j(t)\, f_{\alpha,j}(x,t)), \\ |\alpha| \leqq 2(r-j)\, m, \quad 0 \leqq j \leqq r, \end{cases} \tag{9.12}$$

where

$$f_{\alpha,j} \in L^2(Q). \tag{9.13}$$

Proof. By the definition of $\Xi^{2rm,r}(Q)$ and the Hahn-Banach theorem, every continuous linear form on this space may be written

$$M(\varphi) = \sum_{\alpha,j} \int_Q \varrho^{|\alpha|}\, d^j\, D_t^j\, D_x^\alpha \varphi \cdot g_{\alpha,j}\, dx\, dt, \quad g_{\alpha,j} \in L^2(Q).$$

By Proposition 9.2, M is defined by its values for $\varphi \in \mathscr{D}(Q)$, from which (9.12) follows (setting $f_{\alpha,j} = (-1)^{\alpha+j} g_{\alpha,j}$). $\square$

9.3 Choice of f. The Space $D_P^{-(r-1)}(Q)$

From Propositions 9.1 and 9.2 we obtain:

Proposition 9.4. *For* f *given in* $\Xi^{-2rm,-r}(Q)$,

$$v \to \langle f, \bar{v} \rangle \tag{9.14}$$

is a continuous antilinear form on $X^r(Q)$. $\square$

In (8.8), we shall take $\langle f, \bar{v}\rangle$ under the preceding conditions. Thus we shall obtain a solution $u \in H^{-2(r-1)m,-(r-1)}(Q)$ such that $A u + u' = f$, therefore

$$P u = A u + u' \in \Xi^{-2rm,-r}(Q),$$

which leads to the *following definition:*

$$D_P^{-(r-1)}(Q) = \{u \mid u \in H^{-2(r-1)m,-(r-1)}(Q), \quad P u \in \Xi^{-2rm,-r}(Q)\} \tag{9.15}$$

$$(r \geqq 1, \quad 2r\,m \quad \text{and} \quad r \neq \text{integer} + \tfrac{1}{2}).$$

Provided with the *norm of the graph*

$$\|u\|_{D_P^{-(r-1)}(Q)} = \left(\|u\|^2_{H^{-2(r-1)m,-(r-1)}(Q)} + \|P u\|^2_{\Xi^{-2rm,-r}(Q)}\right)^{1/2} \tag{9.16}$$

$D_P^{-(r-1)}(Q)$ is a Hilbert space. □

We shall now obtain trace theorems for $D_P^{-(r-1)}(Q)$, which will show how (8.6)–(8.8) "contain" (8.10)–(8.11).

We shall follow the methods already used in this book (in particular, see Section 6 of Chapter 2). But we call attention to the fact that we shall not obtain "optimal" results as we did for the elliptic case. Indeed, as will be explained in detail in Section 12.3, we can *globally*, by extension by continuity define the operator $\sigma u = \{\vec{B}_j u, u(x, 0)\}$ for the u's in $D_P^{-(r-1)}(Q)$; but we face rather substantial technical difficulties if we seek to "separate" the operators $u(x, 0)$ on Ω and $\vec{B} u$ on Σ in σu and to obtain the "optimal" spaces for $u(x, 0)$ and $\vec{B} u$ separately. Nevertheless, we believe that the results we shall obtain are already satisfactory from the point of view of applications. □

10. Trace Theorems for the Spaces $D_P^{-(r-1)}(Q)$, $r \geqq 1$

10.1 Density Theorem

Theorem 10.1. *Under the hypotheses* (1.1), (1.2), (1.3) *of Section* 1.1 *and if for every* $\theta \in \left[-\frac{\pi}{2}, \frac{\pi}{2}\right]$ *and* $t_0 \in [0, T]$ *the operator* $A\left(x, t_0, \frac{\partial}{\partial x}\right) + (-1)^m e^{i\theta} D_y^{2m}$ *is properly elliptic in* $\bar{\Omega} \times \mathbf{R}$, *the space* $\mathcal{D}(\bar{Q})$ *is dense in* $D_P^{-(r-1)}(Q)$ *for every* $r \geqq 1$ *with* $2r\,m$ *and* $r \neq$ *integer* $+ \frac{1}{2}$.

Proof. 1) Let $u \to N(u)$ be a continuous antilinear form on $D_P^{-(r-1)}(Q)$; according to (9.15), (9.16) and the Hahn-Banach theorem, we may write:

$$N(u) = \langle \psi_0, \bar{u}\rangle + \langle \psi_1, \overline{Pu}\rangle, \tag{10.1}$$

where

$$\psi_0 \in H^{2(r-1)m, r-1}_{0,0}(Q),$$
$$\psi_1 \in \Xi^{2rm, r}(Q),$$

and where the first (resp. second) bracket denotes the duality between

$$H^{2(r-1)m, r-1}_{0,0}(Q) \quad \text{and} \quad H^{-2(r-1)m, -(r-1)}(Q)$$
$$(\text{resp. } \Xi^{2rm, r}(Q) \quad \text{and} \quad \Xi^{-2rm, -r}(Q)).$$

Assume that

(10.2) $$N(\varphi) = 0 \qquad \forall \varphi \in \mathscr{D}(\bar{Q}).$$

We shall show that this implies

(10.3) $$N(u) = 0 \qquad \forall u \in D_P^{-(r-1)}(Q)$$

(which will prove the theorem).

2) Let $\tilde{\psi}_0$, $\tilde{\psi}_1$ be the extensions of the functions ψ_i to $\mathbf{R}^n_x \times \mathbf{R}_t$ by 0 outside Q. The function ψ_1 belongs, *in particular*, to $L^2(Q)$, therefore $\tilde{\psi}_1 \in L^2(\mathbf{R}^{n+1}_{x,t})$.

Let Φ be arbitrary in $\mathscr{D}(\mathbf{R}^{n+1}_{x,t})$; if φ denotes its restriction to Q, we have:

(10.4) $$\langle \tilde{\psi}_0, \Phi \rangle + \langle \tilde{\psi}_1, \overline{\tilde{P}\,\Phi} \rangle = \langle \psi_0, \bar{\varphi} \rangle + \langle \psi_1, \overline{P\varphi} \rangle$$

where $\tilde{P} = \tilde{A} + \dfrac{\partial}{\partial t}$ is an "extension" of P, $\tilde{A} = \tilde{A}\left(x, t, \dfrac{\partial}{\partial x}\right)$ being an operator defined in $\mathbf{R}^{n+1}_{x,t}$, with coefficients of class C^∞, which coincides with A on Q and is such that the operator

$$\tilde{A}\left(x, t_0, \frac{\partial}{\partial x}\right) + (-1)^m e^{i\theta} D_y^{2m} \quad \text{is, for every} \quad \theta \in \left[\frac{\pi}{2}, \frac{\pi}{2}\right]$$

and every $t_0 \in [-\varepsilon, T + \varepsilon]$, properly elliptic in $\bar{\mathcal{O}} \times \mathbf{R}_y$, where $\mathcal{O}$ is a suitable neighbourhood of $\bar{\Omega}$ and ε is a suitable positive real number; the first (resp. second) bracket on the left-hand side of (10.4) denotes, for example, the duality between $H^{2(r-1)m, r-1}_{0,0}(\mathbf{R}^{n+1}_{x,t})$ and its dual (resp. between $L^2(\mathbf{R}^{n+1}_{x,t})$ and itself). According to (10.1) and (10.2), we have:

$$\langle \tilde{\psi}_0, \Phi \rangle + \langle \tilde{\psi}_1, \overline{\tilde{P}\,\Phi} \rangle = 0,$$

from which we obtain

(10.5) $$\tilde{\psi}_0 + (\tilde{P})^* \tilde{\psi}_1 = 0$$

or

(10.5a) $$\tilde{\psi}_0 + (\tilde{A})^* \tilde{\psi}_1 - \frac{\partial \tilde{\psi}_1}{\partial t} = 0.$$

The function $\tilde{\psi}_1$ belongs to $L^2(\mathcal{O} \times]-\varepsilon, T+\varepsilon[)$, *vanishes outside* Q and satisfies:

$$(10.5\text{b}) \qquad (\tilde{A})^* \tilde{\psi}_1 - \frac{\partial \tilde{\psi}_1}{\partial t} = \psi_0 \quad \text{in} \quad \mathcal{O} \times]-\varepsilon, T+\varepsilon[.$$

But there exists a unique function w (see Theorems 5.3 and 6.2, which apply since the system $\left\{\frac{\partial^j}{\partial \nu^j}\right\}_{j=0}^{m-1}$ covers every properly elliptic operator) such that

$$(10.6) \qquad \begin{cases} (A)^* w - \dfrac{\partial w}{\partial t} = -\tilde{\psi}_0 \quad \text{in} \quad \mathcal{O} \times]-\varepsilon, T+\varepsilon[, \\[2ex] \dfrac{\partial^j w}{\partial \nu^j} = 0 \quad \text{on} \quad \partial\mathcal{O} \times]-\varepsilon, T+\varepsilon[, \quad 0 \leqq j \leqq m-1, \\[2ex] w(x, T+\varepsilon) = 0, \end{cases}$$

where, since (as r and $2rm \neq$ integer $+\frac{1}{2}$) $\tilde{\psi}_0 \in H_{0,0}^{2(r-1)m, r-1}(\mathcal{O} \times]-\varepsilon, T+\varepsilon[)$, we have:

$$(10.7) \qquad w \in H^{2rm, r}(\mathcal{O} \times]-\varepsilon, T+\varepsilon[).$$

We shall show that

$$(10.8) \qquad \tilde{\psi}_1 = w.$$

Indeed, let φ be arbitrary in $\mathcal{D}(Q)$; we introduce v, solution of

$$(10.9) \qquad \begin{cases} \tilde{A} v + v' = \varphi \quad \text{in} \quad \mathcal{O} \\[2ex] \dfrac{\partial^j v}{\partial \nu^j} = 0 \quad \text{on} \quad \partial\mathcal{O} \times]-\varepsilon, T+\varepsilon[, \quad 0 \leqq j \leqq m-1, \\[2ex] v(x, -\varepsilon) = 0. \end{cases}$$

We have

$$(10.10) \qquad \begin{cases} \langle ((\tilde{A})^* - D_t) \psi_1, \bar{v} \rangle = \langle \psi_0, \bar{v} \rangle \quad \text{(by (10.5b))} \\ = \langle \tilde{\psi}_1, \overline{(\tilde{A} + D_t) v} \rangle \quad \text{(since } \psi_1 \text{ has compact support in} \\ \mathcal{O} \times]-\varepsilon, T+\varepsilon[) = \langle \tilde{\psi}_1, \bar{\varphi} \rangle. \end{cases}$$

On the other hand, since w satisfies (10.6) and (10.7) and since v is infinitely differentiable in $\bar{\mathcal{O}} \times [-\varepsilon, T+\varepsilon]$ (since $\bigcap_k H^{k, k/2m}(\mathcal{O} \times]-\varepsilon, T+\varepsilon[) \equiv C^\infty$-functions in $\bar{\mathcal{O}} \times [-\varepsilon, T+\varepsilon]$), we have:

$$\begin{aligned} \langle ((\tilde{A})^* - D_t) w, \bar{v} \rangle &= -\langle \tilde{\psi}_0, \bar{v} \rangle \\ &= \langle w, \overline{(\tilde{A} + D_t) v} \rangle \quad \text{(by Green's formula, Section 1)} \\ &= \langle w, \bar{\varphi} \rangle \end{aligned}$$

which, upon comparison with (10.10) shows that

$$\langle \tilde{\psi}_1, \bar{\varphi} \rangle = \langle w, \bar{\varphi} \rangle \qquad \forall \varphi \in \mathscr{D}(Q);$$

whence (10.8).

Therefore, according to (10.7), $\tilde{\psi}_1 \in H^{2rm,r}(\mathscr{O} \times]-\varepsilon, T+\varepsilon[)$ and since $\tilde{\psi}_1$ vanishes outside Q, it follows that ψ_1 and all its derivatives with well-defined traces vanish on the boundary ∂Q of Q. Therefore

$$\psi_1 \in H^{2rm,r}_{0,0}(\Omega \times]0, T[) \tag{10.11}$$

and (by restriction of (10.5) to Q):

$$\psi_0 + P^* \psi_1 = 0 \quad \text{in} \quad Q. \tag{10.12}$$

3) Now, if $u \in D_P^{-(r-1)}(Q)$, then we have:

$$\langle \overline{P u}, \psi_1 \rangle = \langle \bar{u}, P^* \psi_1 \rangle$$

and therefore

$$N(u) = \langle \psi_0 + P^* \psi_1, \bar{u} \rangle;$$

from which we obtain (10.3) with the help of (10.12), which proves the theorem. □

10.2 Trace Theorem on Σ

Theorem 10.2. *Under the hypotheses of Theorem* 10.1 *and* (1.6) *of Section* 1.1, *the mapping $u \to B_j u$ of $\mathscr{D}(\bar{Q}) \to \mathscr{D}(\bar{\Sigma})$ extends by continuity to a continuous linear mapping, still denoted by $u \to B_j u$, of*

$$D_P^{-(r-1)}(Q) \to H^{-2rm+2m-m_j-1/2, -r+(2m-m_j-1/2)/2m}(\Sigma),$$

for every $r \geqq 1$ with $2rm$ and $r \neq$ integer $+\frac{1}{2}$.

Proof. 1) We shall define all $B_j u$ simultaneously, i.e. $B u$ (notation of Section 1). Let us for the moment accept the following lemma without proof.

Lemma 10.1. *There exists a right inverse*

$$\vec{g} = \{g\}_{j=0}^{m-1} \to R(\vec{g}) = v, \tag{10.13}$$

a continuous linear mapping of

$$\prod_{j=0}^{m-1} H^{2rm-(2m-m_j-1/2), r-(2m-m_j-1/2)/2m}_{,0}(\Sigma) \to \mathscr{V} \tag{10.14}$$

where

$$\mathscr{V} = \{v \mid v \in H^{2rm,r}_{,0}(Q), P^* v \in H^{2(r-1)m, r-1}_{0,0}(Q)\} \tag{10.15}$$

such that

$$\vec{C} v = 0, \quad \vec{T} v = \vec{g}. \tag{10.16}$$

2) For given $u \in D_P^{-(r-1)}(Q)$ and given $\vec{g} \in \prod_{j=0}^{m-1}$ (which here denotes the first space in (10.14)), set

$$Z(\vec{g}) = -\langle P\,u, \bar{v}\rangle + \langle u, \overline{P^*\,v}\rangle, \tag{10.17}$$

where the first (resp. second) bracket denotes the duality between $\Xi^{-2rm,-r}(Q)$ and $\Xi^{2rm,r}(Q)$ which contains $H^{2rm,r}(Q)$ (reps. between $H^{-2(r-1)m,-(r-1)}(Q)$ and $H^{2(r-1)m,r-1}_{0,0}(Q)$) and where $v = R(\vec{g})$.

The (antilinear) form $Z(\vec{g})$ is independent of the right-inverse (which of course is *not unique*); indeed, if v_1 is another function of $\mathscr{V}$ (defined by (10.15)) satisfying (10.16), then $v - v_1 = w$ satisfies:

$$\vec{C}\,w = 0, \quad \vec{T}\,w = 0$$

and therefore, $\{\vec{C}, \vec{T}\}$ being a Dirichlet system of order $2m$:

$$\frac{\partial^k w}{\partial \nu^k} = 0 \quad \text{on} \quad \Sigma, \quad 0 \leqq k \leqq 2m-1. \tag{10.18}$$

But

$$P^*\,w \in H^{2(r-1)m,r-1}_{0,0}(Q), \quad \text{i.e.} \quad A^*\,w - \frac{\partial w}{\partial t} \in H^{2(r-1)m,r-1}_{0,0}(Q)$$

and therefore

$$D_x^p\left(A^*\,w - \frac{\partial w}{\partial t}\right) = 0 \quad \text{on} \quad \Sigma, \quad \forall p \quad \text{with} \quad |p| \leqq 2(r-1)\,m - \frac{1}{2}.$$

Writing first that $A^*\,w - \frac{\partial w}{\partial t} = 0$ on Σ and since, by (10.18), $\frac{\partial w}{\partial t} = 0$ on Σ, we see that $A^*\,w = 0$ on Σ, which, thanks to the ellipticity of A^*, implies $\frac{\partial^{2m} w}{\partial \nu^{2m}} = 0$ on Σ. Next, we take $|p| = 1, \ldots$ and we obtain

$$\frac{\partial^k w}{\partial \nu^k} = 0 \quad \text{on} \quad \Sigma \quad \text{for} \quad 0 \leqq k < 2r\,m - \frac{1}{2}, \tag{10.19}$$

by recurrence.

Furthermore, $v^{(j)}(0) = v^{(j)}(T) = 0$ if $0 \leqq j < r - \frac{1}{2}$, with an analogous condition for v_1, therefore for w, which together with (10.19) shows that

$$w \in H^{2rm,r}_{0,0}(Q)^{((1))}. \tag{10.20}$$

((1)) Indeed, we verify, as in Chapter 1, Section 2, that if w satisfies (10.19) and

$$w^{(j)}(0) = w^{(j)}(T) = 0, \quad 0 \leqq j < r - \tfrac{1}{2},$$

then w belongs to the closure of $\mathscr{D}(Q)$ in $H^{2rm,r}(Q)$, i. e. in $H^{2rm,r}_{0,0}(Q)$.

Then
$$\langle u, \overline{P^* w}\rangle = \langle P u, \bar{w}\rangle,$$
which shows that
$$-\langle P u, \bar{v}\rangle + \langle u, \overline{P^* v}\rangle = -\langle Pu, \bar{v}_1\rangle + \langle u, \overline{P^* v_1}\rangle;$$
whence our assertion.

It follows from Lemma 10.1 that the form $\vec{g} \to Z(\vec{g})$ is continuous on $\prod_{j=0}^{m-1}$, therefore it may be written

$$Z(\vec{g}) = \langle \vec{\beta}_u, \bar{\vec{g}}\rangle, \quad \vec{\beta}_u \in \left(\prod_{j=0}^{m-1}\right)'. \tag{10.21}$$

In this way, we define a linear mapping
$$u \to \vec{\beta}_u \quad \text{of} \quad D_P^{-(r-1)}(Q) \to \left(\prod_{j=0}^{m-1}\right)',$$
which is *continuous*: if $u \to 0$ in $D_P^{-(r-1)}(Q)$ and if $\vec{g}$ belongs to a bounded set of $\prod_{j=0}^{m-1}$, then $R(\vec{g})$ belongs to a bounded set of $\mathscr{V}$, so that, according to the definition (10.17), $Z(\vec{g}) \to 0$ uniformly in $\vec{g}$ and therefore $\vec{\beta}_u \to 0$ in $\left(\prod_{j=0}^{m-1}\right)'$.

3) Now, *assume that* $u \in \mathscr{D}(\bar{Q})$. Then Green's formula (Section 1), which we are now *allowed to use*, shows that

$$Z(\vec{g}) = (\vec{B} u, \vec{T} v)_\Sigma = \langle \vec{B} u, \bar{\vec{g}}\rangle$$
and therefore:

$$\begin{cases} \text{for} \quad u \in \mathscr{D}(\bar{Q}), \text{ we have } \langle \vec{\beta}_u - \vec{B}_u u, \bar{\vec{g}}\rangle = 0 \\ \forall \vec{g} \in \prod_{j=0}^{m-1}, \quad \text{therefore} \quad \vec{\beta}_u = \vec{B} u. \end{cases} \tag{10.22}$$

This proves the theorem if we can prove Lemma 10.1. □

Proof of Lemma 10.1. Let us write all boundary conditions to be satisfied by v.

First, we must have (10.16), which, since $\{\vec{C}, \vec{T}\}$ is a Dirichlet system, is *equivalent* to (see Chapter 2, Section 2):

$$\frac{\partial^k v}{\partial \nu^k} = h_k, \quad 0 \leqq k \leqq 2m - 1, \tag{10.23}$$
with

$$h_k \text{ given in } H^{2rm-(k+1/2),r-(k+1/2)/2m}_{,0}(\Sigma). \tag{10.24}$$

Next, we must have
$$v \in H^{2rm,r}_{,0}(Q) \quad \text{and} \quad P^* v = A^* v - v' \in H^{2(r-1)m,r-1}_{0,0}(Q),$$

which entails two types of boundary conditions:

(i) *on* Σ:

$$\frac{\partial^k A^* v}{\partial \nu^k} = \frac{\partial^k v'}{\partial \nu^k}, \quad 0 \leqq k < 2(r-1)m - \frac{1}{2}; \tag{10.25}$$

(ii) *on* Ω_0 *and* Ω_T:

$$\frac{\partial^j v}{\partial t^j}(x, 0) = \frac{\partial^j v}{\partial t^j}(x, T) = 0, \quad 0 \leqq j < r - \frac{1}{2}. \tag{10.26}$$

Apply (10.25) with $k = 0$; then $v'|_\Sigma = h_0'$, therefore

$$A^* v|_\Sigma = h_0' \in H_{,0}^{2rm-1/2, r-(2m+1/2)/2m}(\Sigma).$$

and this together with (10.23) yields

$$\frac{\partial^{2m} v}{\partial \nu^{2m}} = h_{2m} \in H_{,0}^{2rm-(2m+1/2), r-(2m+1/2)/2m}(\Sigma).$$

By induction, assume that (10.25) implies

$$\frac{\partial^{2m+k} v}{\partial \nu^{2m+k}} = h_{2m+k} \in H_{,0}^{2rm-(2m+k+1/2), r-(2m+k+1/2)/2m}(\Sigma) \tag{10.27}$$

for $k \leqq k_0 - 1$; then, for $k = k_0$, (10.25) yields

$$\frac{\partial^{k_0} A^* v}{\partial \nu^{k_0}} = \left.\frac{\partial^{k_0} v'}{\partial \nu^{k_0}}\right|_\Sigma \in H_{,0}^{2rm-(k_0+1/2), r-(2m+k_0+1/2)/2m}(\Sigma)$$

whence (10.27) for $k = k_0$.

Finally, *the boundary conditions are equivalent to*

$$\begin{cases} \dfrac{\partial^k v}{\partial \nu^k} = h_k \text{ given in } H_{,0}^{2rm-(k+1/2), r-(k+1/2)/2m}(\Sigma), \\ 0 \leqq k < 2rm - \frac{1}{2} \end{cases} \tag{10.28}$$

and to (10.26) (note that we do not have global compatibility conditions, since $2rm$ and $r \neq$ integer $+ \frac{1}{2}$).

Remark 10.1. Of course, we define $S_j u$ for $u \in D_P^{-(r-1)}(Q)$ in the same way, we have:

$$\begin{cases} \textit{the mapping } u \to S_j u \textit{ of } \mathscr{D}(\bar{Q}) \to \mathscr{D}(\bar{\Sigma}) \textit{ extends by continuity} \\ \textit{to a continuous linear mapping, still denoted by } u \to S_j u, \textit{ of} \\ D_P^{-(r-1)}(Q) \to \\ \to H^{-2(r-1)m-(\text{order } S_j)-1/2, -(r-1)-(\text{order } S_j+1/2)/2m}(\Sigma). \end{cases} \tag{10.29}$$

It follows that

$$(10.30)\quad \begin{cases} \textit{the mapping } u \to \dfrac{\partial^k u}{\partial \nu^k} \textit{ of } \mathcal{D}(\bar{Q}) \to \mathcal{D}(\bar{\Sigma}) \textit{ extends by continuity} \\ \textit{to a continuous linear mapping, still denoted by } u \to \dfrac{\partial^k u}{\partial \nu^k}, \textit{ of} \\ D_P^{-(r-1)}(Q) \to H^{-2(r-1)m-k-1/2,-(r-1)-(k+1/2)/2m}(\Sigma). \ \square \end{cases}$$

Remark 10.2. Every distributions $u \in H^{-\infty}(Q) = \bigcup_{k=0}^{\infty} H^{-k}(Q)$ belongs to a space $H^{-2(r-1)m,-(r-1)}(Q)$ for a suitable integer r. Theorem 10.2 therefore implies that, *for every distribution u in $H^{-\infty}(Q)$ such that $Pu \in \bigcup_r \Xi^{-2rm,r}(Q)$, we may define $B_j u$ on Σ as a distribution in $H^{-\infty}(\Sigma)$.*

In a *very* approximate manner, we may say that we shall see in Volume 3 how this can be extended when u is an "arbitrary distribution" on Q such that Pu" does not grow too rapidly" in the neighbourhood of Σ; then we can still define $B_j u$ on Σ, *but they are no longer distributions in the usual sense.* $\square$

Remark 10.3. We now have the following *partial Green's formula* (see also Section 12.3) at our disposal:

$$\langle P u, \bar{v}\rangle - \langle u, \overline{P^* v}\rangle = -\langle \vec{B} u, \bar{\vec{T}} v\rangle$$

for

$$u \in D_P^{-(r-1)}(Q), \quad v \in H^{2rm,r}_{,0}(Q), \quad P^* v \in H^{2(r-1),r-1}_{0,0}(Q)$$

$$C_j v = 0, \quad j = 0, \ldots, m-1. \ \square$$

10.3 Continuity of the Trace on Surfaces Neighbouring Σ

Let Γ_ϱ be a family of surfaces which are "parallel" to Γ and "tend towards Γ" as $\varrho \to 0$ (see Chapter 2, Section 8.1).

We assume that a correspondence

$$(10.31)\qquad x \to \psi(x;\varrho) \quad \text{of} \quad \Gamma_\varrho \to \Gamma,$$

which is infinitely differentiable both ways, has been established.

Let

$$(10.32)\qquad \Sigma_\varrho = \Gamma_\varrho \times]0, T[.$$

We assume, as we are allowed to do, that the coefficients of the operators $\{B_j\}$, $\{C_j\}$, $\{S_j\}$ and $\{T_j\}$ are defined and infinitely differentiable not only on $\bar{\Sigma}$ but in $\bar{Q} - Q_{\varrho_0}$, so that the systems $\{B_j\}$, $\{C_j\}$, $\{S_j\}$ and $\{T_j\}$ are normal on Γ_ϱ for $0 \leqq \varrho \leqq \varrho_0$, $0 \leqq t \leqq T$. We introduce the subspace of $D_P^{-(r-1)}(Q)$ defined by

$$(10.33)\qquad \tilde{D}_P^{-(r-1)}(Q) = \{u \mid u \in H^{-2(r-1)m,-(r-1)}(Q), \quad Pu \in L^2(Q)\}$$

and in analogous fashion

(10.34) $\tilde{D}_P^{-(r-1)}(Q_\varrho) = \{u \mid u \in H^{-2(r-1)m,-(r-1)}(Q_\varrho), \quad P u \in L^2(Q_\varrho)\}$,

with the norms of the graph. We note that the restrictions to Q_ϱ of the u's in $\tilde{D}_P^{-(r-1)}(Q)$ belong to $\tilde{D}_P^{-(r-1)}(Q_\varrho)$. By the same type of proof as in Sections 10.1 and 10.2, we can show that $\mathscr{D}(\bar{Q})$ (resp. $\mathscr{D}(\bar{Q}_\varrho)$) is dense in $\tilde{D}_P^{-(r-1)}(Q)$ (resp. $\tilde{D}_P^{-(r-1)}(Q_\varrho)$) and that we have a trace theorem for the u's in $\tilde{D}_P^{-(r-1)}(Q)$ (resp. $\tilde{D}_P^{-(r-1)}(Q_\varrho)$) which is the analogue of Theorem 10.2. Therefore, we can define $B_j u|_{\Sigma_\varrho}$ and $B_j u|_\Sigma$ for the u's in $\tilde{D}_P^{-(r-1)}(Q)$.

With the aid of transformation (10.31), we can – tranfer of structure – *define the image of* $B_j u|_{\Sigma_\varrho}$ on Σ; we set:

(10.35) $B_j^{(\varrho)} u =$ image of $B_j u|_{\Sigma_\varrho}$ under (10.31).

By the trace theorem (and the fact that (10.31) is infinitely differentiable both ways) we have:

(10.36) $B_j^{(\varrho)} u \in H^{-2(r-1)m-(m_j+1/2),-(r-1)-(m_j+1/2)/2m}(\Sigma)$.

Theorem 10.3. *Hypotheses and notation of Theorem* 10.2. *If* $u \in \tilde{D}_P^{-(r-1)}(Q)$ *and the operators* $B_j^{(\varrho)}$ *are defined by* (10.35), *then*

(10.37) $B_j^{(\varrho)} u \to B_j u$ *in* $H^{-2(r-1)m-(m_j+1/2),-(r-1)-(m_j+1/2)/2m}(\Sigma)$

as $\varrho \to 0$.

Proof. If $u \in \mathscr{D}(\bar{Q})$, then (10.37) is obvious – with convergence in $\mathscr{D}(\Sigma)$. Since $\mathscr{D}(\bar{Q})$ is dense in $\tilde{D}_P^{-(r-1)}(Q)$, it is sufficient to show that

(10.38) $$\begin{cases} B_j^{(\varrho)} \text{ remains in a bounded set of} \\ \mathscr{L}(\tilde{D}_P^{-(r-1)}(Q); H^{-2(r-1)m-(m_j+1/2),-(r-1)-(m_j+1/2)/2m}(\Sigma)), \\ \text{as } \varrho \to 0. \end{cases}$$

Set

$$\prod_j^{(\varrho)} = \prod_{j=0}^{m-1} H^{-2(r-1)m-(m_j+1/2),-(r-1)-(m_j+1/2)/2m}(\Sigma_\varrho)$$

and (compare with (10.15)):

$$\mathscr{V}_\varrho = \{v \mid v \in H^{2rm,r}_{,0}(Q_\varrho),\ P^* v \in H^{2(r-1)m,r-1}_{0,0}(Q_\varrho)\}.$$

For g fixed in

$$\prod_{j=0}^{m-1} H^{-2(r-1)m-(m_j+1/2),-(r-1)-(m_j+1/2)/2m}(\Sigma),$$

we have to show that $|\langle B^{(\varrho)} u, \bar{g}\rangle| \leqq$ constant (independent of ϱ).

After a transfer of structure by (10.31), this amounts to showing that

$$|\langle \vec{B} u|_{\Sigma_\varrho}, \bar{g}_\varrho\rangle| \leqq \text{constant},$$

where (in particular) g_ϱ belongs to a bounded set of $\prod_j^{(\varrho)}$. But (by the definition of $\vec{B}\, u\,|_{\Sigma_\varrho}$):

$$\langle \vec{B}\, u\,|_{\Sigma_\varrho}, \bar{g}_\varrho\rangle = \langle u, \overline{P^*\, v_\varrho}\rangle_{Q_\varrho} - \langle P\, u, v_\varrho\rangle_{Q_\varrho},$$

where

$$v_\varrho = R_\varrho(g_\varrho)$$

and R_ϱ = right-inverse analogous to the one in Lemma 6.1 but passing from Σ_ϱ to Q_ϱ (instead of Σ to Q).

It suffices now to verify that we can choose R_ϱ in such a way that $R_\varrho(g_\varrho)$ remains in a ball of $\mathscr{V}_\varrho$ with radius independent of ϱ, as $\varrho \to 0$. However, this follows from the proof of Lemma 10.1 and from Theorem 2.2, since all local maps take place with functions which, together with all their derivatives, are bounded in ϱ. □

Remark 10.4. Assume that u is given with

$$u \in H^{-\infty}(Q), \quad P\, u = 0. \tag{10.39}$$

Then (see also Remark 10.2) $u \in H^{-2(r-1)m, -(r-1)}(Q)$ for a suitable r and on the other hand, since P *is hypoelliptic* (see S. Mizohata [2]; we shall prove this in Volume 3), we have:

$$u \in \mathscr{E}(Q) = C^\infty(Q). \tag{10.40}$$

Therefore

$$B_j\, u\,|_{\Sigma_\varrho} \in \mathscr{E}(\Sigma_\varrho). \tag{10.41}$$

Therefore in this case

$$B_j^{(\varrho)}\, u \in H^{-2(r-1)m-(m_j+1/2), -(r-1)-(m_j+1/2)/2m}(\Sigma) \cap \mathscr{E}(\Sigma), \tag{10.42}$$

(*but this does not alter the nature of the convergence in* (10.37)). □

10.4 Trace Theorem on Ω_0

Theorem 10.4. *Under the hypotheses of Theorem* 10.2, *the mapping* $u \to u(x, 0)$ *of* $\mathscr{D}(\bar{Q}) \to \mathscr{D}(\bar{\Omega})$ *extends by continuity to a continuous linear mapping, still denoted by* $u \to u(x, 0)$, *of* $D_P^{-(r-1)}(Q) \to H^{-2(r-1/2)m}(\Omega)$, *for all* $r \geqq 1$ *with* $2r\, m$ *and* $r \neq$ *integer* $+ \frac{1}{2}$.

Proof. Let φ be given in $H_0^{2(r-1/2)m}(\Omega)$. For the moment we assume

Lemma 10.2. *There exists a right-inverse*

$$\varphi \to R_1(\varphi) = v,$$

which is a continuous linear mapping of $H_0^{2(r-1/2)m}(\Omega) \to \mathscr{V}_1$, *where*

$$\mathscr{V}_1 = \{v \mid v \in H^{2rm, r}(Q),\, P^*\, v \in H_{0,0}^{2(r-1)m, r-1}(Q)\}, \tag{10.43}$$

such that

$$\vec{C}\, v_1 = 0, \quad \vec{T}\, v_1 = 0 \tag{10.44}$$

and

$$v_1(x, T) = 0, \quad v_1(x, 0) = \varphi(x). \tag{10.45}$$

Then if we set

$$Z_1(\varphi) = -\langle P\,u, \bar{v}_1\rangle + \langle u, \overline{P^*\, v_1}\rangle, \tag{10.46}$$

we note (as in the proof of Theorem 10.2) that Z_1 is independent of the choice of the right-inverse (satisfying (10.44) and (10.45)), therefore depends only on φ. The antilinear form $\varphi \to Z_1(\varphi)$ is continuous on $H_0^{2(r-1/2)m}(\Omega)$ and therefore of the form

$$Z_1(\varphi) = \langle \tau_u, \overline{\varphi(x, 0)}\rangle, \quad \tau_u \in H^{-2(r-1/2)m}(\Omega). \tag{10.47}$$

The linear mapping $u \to \tau_u$ maps $D_P^{-(r-1)}(Q) \to H^{-2(r-1/2)m}(\Omega)$ *continuously*.

If $u \in \mathscr{D}(\bar{Q})$, then (we have done all that is necessary for this)

$$\tau_u = u(x, 0),$$

which proves the theorem if we can prove Lemma 10.2. □

Proof of Lemma 10.2. First of all, conditions (10.44) are equivalent to

$$\frac{\partial^k v}{\partial \nu^k} = 0 \quad \text{on} \quad \Sigma, \quad 0 \leqq k \leqq 2\,m-1, \tag{10.48}$$

and (see the proof of Theorem 10.2) the other boundary conditions will be satisfied if we (bluntly) take

$$\frac{\partial^k v}{\partial \nu^k} = 0 \quad \text{on} \quad \Sigma, \quad 0 \leqq k < 2r\,m - \frac{1}{2}. \tag{10.49}$$

Choose v_1 such that

$$v_1^{(j)}(x, T) = 0, \quad 0 \leqq j < r - \tfrac{1}{2}. \tag{10.50}$$

Next, we must have

$$v_1(x, 0) = \varphi(x),$$
$$P^*\, v_1(x, 0) = 0,$$

i.e.

$$v_1'(x, 0) = A^*\,\varphi \in H_0^{2(r-1-1/2)m}(\Omega)$$

and so on, whence

$$\begin{cases} v_1^{(j)}(x, 0) = \varphi_j \ \text{ given in } \ H_0^{2(r-j-1/2)m}(\Omega), \\ 0 \leqq j < r - \tfrac{1}{2}, \quad \varphi_0 = \varphi. \end{cases} \tag{10.51}$$

According to the trace theorem, Chapter 1, Sections 3 and 4 and according to Chapter 1, Section 11, there exists a continuous linear mapping

$$\{\varphi_j\}_{j=0}^{r-1} \to v, \quad \text{of} \quad \prod_{j=0}^{r-1} H_0^{2(r-j-1/2)m}(\Omega) \to H^{2rm,r}(Q),$$

which satisfies (10.49), (10.50) and (10.51), whence the lemma. □

Remark 10.5 (Partial Green's formula). We now have the following "partial Green's formula" at our disposal (compare with Remark 10.3):

$$(10.52)\quad \begin{cases} \langle P\,u, \bar{v}\rangle - \langle u, \overline{P^* v}\rangle = -\langle u(x,0), \overline{v(x,0)}\rangle, \\ u \in D_P^{-(r-1)}(Q), \quad v \in H^{2rm,r}(Q), \quad P^* v \in H_{0,0}^{2(r-1)m,r-1}(Q), \\ v(x,T) = 0, \\ \vec{C}\,v = \vec{T}\,v = 0. \quad \square \end{cases}$$

10.5 Continuity of the Trace on Sections Neighbouring Ω

Denote by Ω_{t_0} the section

$$\Omega_{t_0} = Q \cap \{t = t_0\}, \qquad 0 < t_0 < T.$$

By *restriction* to $Q_{t_0} = \Omega_{t_0} \times \,]t_0, T[$, if $u \in \tilde{D}_P^{-(r-1)}(Q)$, we have $u \in \tilde{D}_P^{-(r-1)}(Q_{t_0})$ (definition analogous to (10.34)) and therefore, by the analogue of Theorem 10.4 for $\tilde{D}_P^{-(r-1)}(Q)$, we may define

$$u(x, t_0) \in H^{-2(r-1/2)m}(\Omega_{t_0}) \approx H^{-2(r-1/2)m}(\Omega).$$

Then, with the same type of method as for Theorem 10.3, we can prove

Theorem 10.5. *Hypotheses and notation of Theorem* 10.4. *For* u *given in* $\tilde{D}_P^{-(r-1)}(Q)$, *we have*

$$(10.53)\quad u(x, t_0) \to u(x, 0) \quad in \quad H^{-2(r-1/2)m}(\Omega) \quad as \quad t_0 \to 0.$$

11. Choice of g_j and u_0. The Spaces $H^{2\alpha m}\,\Xi^{\alpha}(\Sigma)$

11.1 The Spaces $H^{2\alpha m}\,\Xi^{\alpha}(\Sigma)$

We start with the case *integer* α ($\geqq 0$).

Let $d(t)$ be the function defined in (9.3). We define:

$$(11.1)\quad H^{2\alpha m}\,\Xi^{\alpha}(\Sigma) = \{v \mid d^j(t)\, v^{(j)} \in L^2(0, T; H^{2(\alpha - j/\alpha)m}(\Gamma)),$$
$$0 \leqq j \leqq \alpha\};$$

provided with the norm

$$(11.2)\qquad \left(\sum_{j=0}^{\alpha}\int_0^T \|d^j(t)\, v^{(j)}(t)\|^2_{H^{2(\alpha-j/\alpha)m}(\Gamma)}\, dt\right)^{1/2}$$

this space is a *Hilbert space.*

For the case *non-integer* α we use interpolation and intermediate spaces; for $\alpha = k + 1 - \theta$ and $0 < \theta < 1$, we define

$$(11.3)\qquad H^{2\alpha m}\,\Xi^{\alpha}(\Sigma) = [H^{2(k+1)m}\,\Xi^{k+1}(\Sigma), H^{2km}\,\Xi^{k}(\Sigma)]_{\theta}.$$

Proposition 11.1. *The space $\mathcal{D}(\Sigma)$ is dense in $H^{2\alpha m}\,\Xi^{\alpha}(\Sigma)$.*

Proof. 1) Since X is dense in $[X, Y]_{\theta}$, it is sufficient to prove the proposition for *integer* α.

2) Introduce $\varphi_{\varepsilon}(t)$ as in the proof of Proposition 9.1 and, for u given in $H^{2\alpha m}\,\Xi^{\alpha}(\Sigma)$, introduce the functions

$$(11.4)\qquad u_{\varepsilon} = \varphi_{\varepsilon}(t)\, u.$$

It is sufficient to show (Γ being a compact variety) that

$$u_{\varepsilon} \to u \quad \text{in} \quad H^{2\alpha m}\,\Xi^{\alpha}(\Sigma) \quad \text{as} \quad \varepsilon \to 0,$$

therefore that

$$d^j(t)\, u_{\varepsilon}^{(j)} \to d^j(t)\, u^{(j)} \quad \text{in} \quad L^2(0, T; H^{2(\alpha-j/\alpha)m}(\Gamma)), \quad 0 \leqq j \leqq \alpha,$$

and therefore, replacing u_{ε} by its value, that

$$(11.5)\qquad d^j(t)\, \varphi_{\varepsilon}^{(k)}\, u^{(j-k)}(t) \to 0 \quad \text{in} \quad L^2(0, T; H^{2(\alpha-j/\alpha)m}(\Gamma)), \quad 1 \leqq k \leqq j;$$

but this can be written

$$(d^k(t)\, \varphi_{\varepsilon}^{(k)}(t))\, d^{j-k}(t)\, u^{(j-k)}(t),$$

which tends towards zero in $L^2(0, T; H^{2[\alpha-(j-k)/\alpha]/m}(\Gamma))$ and therefore in $L^2(0, T; H^{2(\alpha-j/\alpha)m}(\Gamma))$; whence (11.5) and the proposition. □

Now, we *define*:

$$(11.6)\qquad H^{-2\alpha m}\,\Xi^{-\alpha}(\Sigma) = \text{ dual of } H^{2\alpha m}\,\Xi^{\alpha}(\Sigma), \quad \text{real } \alpha \geqq 0.$$

Thanks to Proposition 11.1, we have:

$$(11.7)\qquad H^{-2\alpha m}\,\Xi^{-\alpha}(\Sigma) \subset \mathcal{D}'(\Sigma).$$

For integer α, the structure of $H^{-2\alpha m}\,\Xi^{-\alpha}(\Sigma)$ is given by:

$$(11.8)\qquad \begin{cases} \text{every distribution } f \in H^{-2\alpha m}\,\Xi^{-\alpha}(\Sigma) \text{ may be represented by} \\ f = \displaystyle\sum_{j=0}^{\alpha} D_t^j(d^j(t)\, f_j), \\ \text{where } f_j \in L^2(0, T; H^{-2(\alpha-j/\alpha)m}(\Gamma)). \quad \square \end{cases}$$

11.2 Choice of g_j

Proposition 11.2. *Let g_j be given with*

$$g_j \in H^{-2(r-1)m-(m_j+1/2)}\, \Xi^{-(r-1)-(m_j+1/2)/2m}(\Sigma), \tag{11.9}$$

for $0 \leqq j \leqq m-1$. Then the antilinear form

$$v \to \sum_{j=0}^{m-1} \langle g_j, \overline{T_j v} \rangle \tag{11.10}$$

is continuous on $X^r(Q)$, $r \geqq 1$, $2rm$ and $r \neq$ integer $+ \frac{1}{2}$.

Proof. If $v \in X^r(Q)$, it follows from Theorem 2.1 and the fact that the order of $T_j = 2m - m_j - 1$ that

$$\begin{aligned} T_j v \in H^{2rm-(2m-m_j-1/2),r-(2m-m_j-1/2)/2m}(\Sigma) &= \\ = H^{2(r-1)m+m_j+1/2,r-1+(m_j+1/2)/2m}(\Sigma) &\subset \\ \subset H^{2(r-1)m+m_j+1/2}\, \Xi^{r-1+(m_j+1/2)/2m}(\Sigma)&, \end{aligned}$$

the mapping $v \to T_j v$ being continuous; whence the proposition. □

11.3 Choice of u_0

Here, we use the spaces $\Xi^{\alpha}(\Omega)$, $\Xi^{-\alpha}(\Omega)$ introduced in Chapter 2, Section 6.

Proposition 11.3. *Let u_0 be given in $\Xi^{-2(r-1/2)m}(\Omega)$. Then the antilinear form*

$$v \to \langle u_0, \overline{v(x,0)} \rangle \tag{11.11}$$

is continuous on $X^r(Q)$, $r \geqq 1$, $2rm$ and $r \neq$ integer $+ \frac{1}{2}$.

Proof. Since $X^r(Q) \subset H^{2rm,r}(Q)$, $v \to v(x, 0)$ is a continuous linear mapping of

$$X^r(Q) \to H^{2rm-m}(\Omega) \subset \Xi^{2(r-1/2)m}(\Omega),$$

from which the result follows.

12. Transposition of the Adjoint Isomorphism of Order r. (II): Results; Existence of Solutions in $H^{2mr,r}(Q)$-Spaces, Real $r \leq 0$

12.1 Final Choice of L

Let f, g_j, and u_0 be given with

$$f \in \Xi^{-2rm,r}(Q) \quad (r \geqq 1, 2rm \text{ and } r \neq \text{integer} + \tfrac{1}{2}), \tag{12.1}$$

$$g_j \in H^{-2(r-1)m-(m_j+1/2)}\, \Xi^{-(r-1)-(m_j+1/2)/2m}(\Sigma), \quad 0 \leqq j \leqq m-1, \tag{12.2}$$

$$u_0 \in \Xi^{-2(r-1/2)m}(\Omega). \tag{12.3}$$

Then, according to Propositions 9.4, 11.2 and 11.3 the antilinear form

(12.4) $$L(v) = \langle f, \bar{v}\rangle + \sum_{j=0}^{m-1} \langle g_j, \overline{T_j v}\rangle + \langle u_0, \overline{v(x,0)}\rangle$$

is continuous on $X^r(Q)$. This is the form L which we choose for (8.1). In this manner, we have justified Section 8.4. □

Remark 12.1. The choice of L we have just made *is not* the most general one possible; see Sections 12.3 and 13. □

12.2 Results

Theorem 12.1. *Hypotheses of Section* 1.1; *r is real and $r \geqq 1$, with $2rm$ and $r \neq$ integer $+ \frac{1}{2}$. Let f, g_j, u_0 be given with* (12.1), (12.2) *and* (12.3). *There exists a unique distribution $u \in D_P^{-(r-1)}(Q)$ satisfying*

(12.5) $$\langle u, \overline{P^* v}\rangle = L(v) \quad \forall v \in X^r(Q) \quad \text{(see Definition 2.1)},$$

where L is given by (12.4). *Furthermore*

(12.6) $$\{f, g_j, u_0\} \to u$$

is a continuous linear mapping of

$$\Xi^{-2rm-r}(Q) \times \prod_{j=0}^{m-1} H^{-2(r-1)m-(m_j+1/2)}\, \Xi^{-(r-1)-(m_j+1/2)/2m}(\Sigma) \times$$
$$\times\, \Xi^{-2(r-1/2)m}(\Omega) \to D_P^{-(r-1)}(Q).$$

The distribution u satisfies

(12.7) $$A u + u' = f \quad in \quad Q,$$

(12.8) $$B_j u = g_j \quad on \quad \Sigma, \quad 0 \leqq j \leqq m-1$$

($B_j u$ *taken in the sense of Theorem* 10.2),

(12.9) $$u(x,0) = u_0 \quad in \quad \Omega$$

($u(x,0)$ *taken in the sense of Theorem* 10.4).

Remark 12.2. We must be very careful about the following point: the distribution u *is not* necessarily *unique* in satisfying (12.7), (12.8) and (12.9); uniqueness occurs only if the boundary conditions (12.8) and (12.9) are interpreted in a *stronger* sense than in Theorems 10.2 and 10.4, as we shall indicate in Section 12.3. □

Proof. 1) By (8.1), there exists a unique u in

$$(H^{2(r-1)m, r-1}_{0,0}(Q))' = H^{-2(r-1)m, -(r-1)}(Q)$$

such that

$$(12.10) \qquad \langle u, \overline{A^* v - \bar{v}'} \rangle = L(v) \quad \forall v \in X^r(Q),$$

L given by (12.4).

Furthermore u depends continuously on L, which will prove the continuity of (12.6).

2) In particular, we have (12.10) for $v = \varphi \in \mathcal{D}(Q)$, therefore (12.7), which entails that $u \in D_P^{-(r-1)}(Q)$.

3) Now we take

$$v \in H^{2rm, r}_{\ ,0}(Q), \quad P^* v \in H^{2(r-1)m, r-1}_{0,0}(Q), \quad \vec{C} v = 0$$

(thus $v \in X^r(Q)$) and apply Remark 10.2; we have:

$$(12.11) \qquad \langle P u, \bar{v} \rangle - \langle u, \overline{P^* v} \rangle = -\langle \vec{B} u, \overline{\vec{T} v} \rangle;$$

but

$$\langle P u, \bar{v} \rangle = \langle f, \bar{v} \rangle$$

and

$$\langle u, \overline{P^* v} \rangle = L(v) = \langle f, \bar{v} \rangle + \langle \vec{g}, \overline{\vec{T} v} \rangle \quad (\text{since } v(x, 0) = 0),$$

so that (12.11) becomes

$$(12.12) \qquad \langle \vec{g}, \overline{\vec{T} v} \rangle = \langle \vec{B} u, \overline{\vec{T} v} \rangle.$$

Since, according to Lemma 10.1, we may take $\vec{T} v$ in (12.11) arbitrary in

$$\prod_{j=0}^{m-1} H^{2rm-(2m-m_j-1/2), r-(2m-m_j-1/2)/2m}_{\ ,0}(\Sigma),$$

(12.8) follows from (12.12).

4) Finally, we take

$$v \in H^{2rm, r}(Q), \quad P^* v \in H^{2(r-1)m, r-1}_{0,0}(Q), \quad v(x, T) = 0, \quad \vec{C} v = \vec{T} v = 0.$$

Then (10.52) yields

$$(12.13) \qquad \langle P u, \bar{v} \rangle - \langle u, \overline{P^* v} \rangle = -\langle u(x, 0), \overline{v(x, 0)} \rangle.$$

But here again

$$\langle P u, \bar{v} \rangle = \langle f, \bar{v} \rangle$$

and

$$\langle u, \overline{P^* v} \rangle = L(v) = \langle f, \bar{v} \rangle + \langle u_0, \overline{v(x, 0)} \rangle \quad (\text{since } \vec{T} v = 0),$$

which, together with (12.13), yields

$$\langle u_0, \overline{v(x, 0)} \rangle = \langle u(x, 0), \overline{v(x, 0)} \rangle,$$

from which, thanks to Lemma 10.2, we obtain (12.9). □

12.3 Complements

We shall complete the statements made in Section 9.3 and in Remarks 10.3, 10.4, 12.1 and 12.2, relative to the fact that the results we have obtained are not the best possible ones. Indeed, for example, in the existence theorem (Theorem 12.1) we have taken the data g_j and u_0 in *smaller* spaces than the ones in which we have given trace theorems (Theorems 10.2 and 10.4) for $B_j u$ and $u(x, 0)$, $u \in D_P^{-(r-1)}(Q)$.

Thus, we have not obtained a *characterization* of traces of u's belonging to $D_P^{-(r-1)}(Q)$, in opposition to what we have been able to do in the elliptic case (see Chapter 2, Section 6, Theorems 6.5 and 6.6).

Let us take a closer look at the difficulties encountered in this problem. As in the elliptic case (Chapter 2, Section 6.2), in order to obtain optimal trace theorems we must study the image $\mathscr{T}(X^r(Q))$ of $X^r(Q)$ under the mapping $\mathscr{T}$ defined by

$$v \to \mathscr{T} v = \{T_0 v, \ldots, T_{m-1} v; v(x, 0)\}, \quad v \in X^r(Q).$$

But, thanks to Theorem 2.3, we see that $\mathscr{T}(X^r(Q))$ coincides with the *subspace* of the product space

$$\prod_{j=0}^{m-1} H^{2(r-1)m+m_j+1/2,\, r-1+(m_j+1/2)/2m}(\Sigma) \times H^{2(r-1/2)m}(\Omega) \tag{12.14}$$

formed by the functions $\{\varphi_0, \ldots, \varphi_{m-1}; \psi\} = \{\vec{\varphi}, \psi\}$ which satisfy the compatibility relations[1]

$$\varphi_j^{(k)}(x, T) = 0, \quad 0 \leqq k < r - \tfrac{3}{2}, \quad x \in \Gamma, \tag{12.15}$$

$$\begin{cases} T_j\left(x, 0, \dfrac{\partial}{\partial x}\right) \psi = \varphi_j(x, 0) \quad \text{on } \Gamma \\ \text{for all } j \text{ such that } 2m - m_j < 2(r - \tfrac{1}{2})\, m + \tfrac{1}{2} \\ \text{(so that } T_j \psi \text{ is well-defined),} \end{cases} \tag{12.16}$$

$$\begin{cases} C_j\left(x, 0, \dfrac{\partial}{\partial x}\right) \psi = 0 \quad \text{on } \Gamma \\ \text{for all } j \text{ such that } 2m - \mu_j < 2(r - \tfrac{1}{2})\, m + \tfrac{1}{2}. \end{cases} \tag{12.17}$$

We denote this subspace by $\mathscr{V}_r$.

According to Theorem 2.3, there exists a right-inverse

$$\{\vec{\varphi}, \psi\} \to \tilde{R}(\vec{\varphi}, \psi) = v \tag{12.18}$$

---

[1] Note that in the case of the Dirichlet problem $\left(B_j = C_j = \dfrac{\partial^j}{\partial \nu^j}\right)$ and if $r = 1$, the compatibility relations are reduced to $\psi \in H_0^m(\Omega)$; see Remark 12.3 on this point.

which is a continuous linear mapping of

$$\mathscr{V}_r \to \{v \mid v \in H^{2rm,r}(Q),\, P^* v \in H^{2(r-1)m,r-1}_{0,0}(Q)\}$$

such that

(12.19) $$v^{(j)}(x, T) = 0, \quad 0 \leqq j < r - \tfrac{3}{2},$$

(12.20) $$\vec{C}\, v = 0, \quad \vec{T}\, v = \vec{\varphi},$$

(12.21) $$v(x, 0) = \psi(x).$$

Thus $v \in X^r(Q)$ and $\tilde{R}$ is a continuous linear right-inverse of the mapping $\mathscr{T}$.

For u given in $D_P^{-(r-1)}(Q)$, we consider the antilinear form

(12.22) $$\{\vec{\varphi}, \psi\} \to Z(\vec{\varphi}, \psi) = -\langle P\, u, \bar{v}\rangle + \langle u, \overline{P^* v}\rangle, \quad v = \tilde{R}(\vec{\varphi}, \psi).$$

As for Theorems 10.2 and 10.4, we verify that $Z(\vec{\varphi}, \psi)$ is independent of the choice of the right-inverse and continuous. It follows that

(12.23) $$Z(\vec{\varphi}, \psi) = \langle \sigma\, u, \overline{\{\vec{\varphi}, \psi\}}\rangle,$$

where

(12.24) $$\sigma\, u \in \mathscr{V}_r',$$

the mapping $u \to \sigma\, u$ being a *continuous* linear mapping of $D_P^{-(r-1)}(Q)$ into this space.

According to the Green's formula of Section 1 and (7.6), we easily see that if $u \in \mathscr{D}(\bar{Q})$, then we can identify $\sigma\, u$ with $\{\vec{B}\, u, u(x, 0)\}$ and therefore, thanks to Theorem 10.1, $\sigma\, u$ is the extension by continuity of the operator $\{\vec{B}\, u, u(x, 0)\}$ from $\mathscr{D}(\bar{Q})$ to $D_P^{-(r-1)}(Q)$.

Thus, we habe obtained a *new trace theorem* for the u's belonging to $D_P^{-(r-1)}(Q)$:

(12.25) *the mapping* $u \to \sigma\, u = \{\vec{B}\, u, u(x, 0)\}$ *of* $\mathscr{D}(\bar{Q})$ *into* $[\mathscr{D}(\Sigma)]^m \times \mathscr{D}(\bar{\Omega})$ *extends by continuity to a continuous linear mapping, still denoted by* $u \to \sigma\, u$, *of* $D_P^{-(r-1)}Q)$ *into* $\mathscr{V}_r'$.

We also have the *Green's formula*:

(12.26) $$\begin{cases} \langle -P\, u, \bar{v}\rangle + \langle u, \overline{P^* v}\rangle = \langle \sigma\, u, \overline{\mathscr{T}\, v}\rangle \\ \forall u \in D_P^{-(r-1)}(Q) \quad \text{and} \quad \forall v \in X^r(Q). \end{cases}$$

Finally, we have

Theorem 12.2. *Under the hypotheses of Theorem* 12.1, *if* f *is given in* $\Xi^{-2rm,-r}(Q)$ *and* g^* *in* $\mathscr{V}_r'$, *then there exists a unique* u *in* $D_P^{-(r-1)}(Q)$

satisfying

(12.27) $$A u + u' = f,$$

(12.28) $$\sigma u = g^*,$$

where (12.27) *is understood in the sense of* $\mathscr{D}'(Q)$ *and* (12.28) *in the sense of* (12.25).

But now the problem is to *separate* the operators $\vec{B} u$ on Σ and $u(x, 0)$ in Ω in σu.

Now, $\mathscr{V}'_r$ is a *quotient space* of

$$\left\{\prod_{j=0}^{m-1} H^{2(r-1)m+m_j+1/2,r-1+(m_j+1/2)/2m}(\Sigma)\right\}' \times (H^{2(r-1/2)m}(\Omega))'$$

which is the dual space of (12.14).

By suitable identifications, this space can be interpreted as a product of spaces of distributions in the usual sense on $\Omega \times \Sigma \times \Gamma$; the introduction of Γ is necessary in order to account for the quotient; therefore for the case of generalized solutions, the boundary data – if we want to interpret them in the usual form – do not appear as in the case of ordinary solutions, see Baiocchi [2] (where this author restricts his arguments to second order operators, but his method is general); see also Lions/Magenes [2].

In order to avoid the technical difficulties inherent to the optimal case, we have, via Theorems 10.2 and 10.4, given meaning to the operators $\vec{B} u$ and $u(x, 0)$ separately in spaces which are in a certain sense *larger* than the restrictions to Σ and Ω of the "optimal" space $\mathscr{V}'_r$. (Note that Sections 10.2 and 10.4 served to insure that the relations (12.15) and (12.16) hold by taking all terms of (12.16) to be *zero*.)

And, in (existence) Theorem 12.1, the data g_j and u_0 belong to spaces *smaller* than the "optimal" spaces. ☐

Remark 12.3. For the Dirichlet problem with $r = 1$, $\mathscr{V}_r$ coincides with the product space

$$\prod_{j=0}^{m-1} H^{j+1/2,(j+1/2)/2m}(\Sigma) \times H_0^m(\Omega)$$

of which the dual is (since $(j + \frac{1}{2})/2m < \frac{1}{2}$, see Chapter 1, Section 11)

$$\prod_{j=0}^{m-1} H^{-j-1/2,-(j+1/2)/2m}(\Sigma) \times H^{-m}(\Omega).$$

Then the "separation" is possible and condition (12.18) can be written in the form

(12.29) $$\gamma_j u = g_j, \quad j = 0, \ldots, m-1,$$

(12.30) $$u(x, 0) = u_0,$$

where $g_j \in H^{-j-1/2,-(j+1/2)/2m}(\Sigma)$, $u_0 \in H^{-m}(\Omega)$ and

$$(12.31)\quad \begin{cases} \textit{the mapping } u \to \{A\,u + u'; \gamma_j u, j = 0, \ldots, m-1; \\ u(x,0)\} \textit{ is an isomorphism of } D_P^0(Q) \textit{ onto} \\ \Xi^{-2m,-1}(Q) \times \prod_{j=0}^{m-1} H^{-j-1/2,-(j+1/2)/2m}(\Sigma) \times H^{-m}(\Omega). \end{cases}$$

Furthermore, for this case, the space $\Xi^{2m,1}(Q)$ can be replaced by the smaller space

$$\{v \mid v \in L^2(0, T; H_0^m(\Omega)), dv' \in L^2(0, T; L^2(\Omega))\},$$

where d is defined in Section 9.1, and then $A\,u + u'$ can be taken in the dual space.

13. State of the Problem. Complements on the Transposition of the Adjoint Isomorphism of Order 1

13.1 State of the Problem

Let us briefly recall Theorems 6.2 and 12.1; the problem

$$(13.1)\quad \begin{cases} A\,u + u' = f, \\ B_j u = g_j, \quad j = 0, \ldots, m-1, \\ u(x,0) = u_0, \end{cases}$$

is solvable with:

$$\text{(i)}\quad \begin{cases} u \in H^{2m(s+1),s+1}(Q), \textit{ if } f \in H^{2ms,s}(Q), \\ g_j \in H^{2m(s+1)-m_j-1/2,\, s+1-(m_j+1/2)/2m}(\Sigma),\ u_0 \in H^{2(s+1/2)m}(\Omega) \\ \textit{and satisfy the compatibility relations; and this for } s \geqq 0, \textit{ with} \\ 2s\,m \textit{ and } s \neq \textit{integer} + \tfrac{1}{2}; \end{cases}$$

$$\text{(ii)}\quad \begin{cases} u \in H^{2m(s+1),s+1}(Q), \textit{ if } f \in \Xi^{2m,s}(Q), \\ g_j \in H^{2m(s+1)-m_j-1/2,\, s+1-(m_j+1/2)/2m}(\Sigma),\ u_0 \in \Xi^{2(s+1/2)m}(\Omega); \\ \textit{and this for } s \leqq -1, \textit{ with } 2s\,m \textit{ and } s \neq \textit{integer} + \tfrac{1}{2}. \end{cases}$$

Note that in (i) and (ii), u is a solution of (13.1) in the sense that $A\,u + u' = f$ is satisfied in $\mathscr{D}'(Q)$ and the conditions $B_j u = g_j$ and $u(x,0) = u_0$ are satisfied in the sense of Theorem 2.1 for (i) and Theorems 10.2 and 10.4 for (ii).

In summary form, we say that *the data are of* "regularity s", $s \in \mathbf{R}$ (s of arbitrary sign), if $f \in H^{2ms,s}(Q)$ or $\Xi^{2ms,s}(Q)$, and g_j and u_0 are "suitably" associated.

Then (i) and (ii) mean this:

$$\begin{cases} \textit{if the data are of "regularity } s\textit{", with } s \geqq 0 \textit{ or } s \leqq -1, \textit{ then} \\ \textit{the solution } u \textit{ belongs to } H^{2m(s+1), s+1}(Q). \end{cases}$$

Therefore it remains to study the case *in which the data are of "regularity* s*", with* $-1 < s < 0$.

Following the "plan" of this book, it would be natural to interpolate between (i) and (ii), in particular between the case $s = 0$ of (i) and $s = -1$ of (ii). But it must be pointed out that we are not in the same situation as, for example, in the elliptic case studied in Section 7 of Chapter 2, because, whereas Theorem 6.2 yields an "optimal" isomorphism theorem for $s \geqq 0$, the situation is different for $s \leqq -1$ since the result of (ii) is not an "optimal" isomorphism theorem.

Furthermore, interpolation between the spaces of data which appear in (i) and (ii) involves considerable technical difficulties.

Thus we shall reconsider the problem without using (ii); we shall start directly with the transposition of the adjoint isomorphism of order 1, according to the notation of Section 7 and without making any simplification; we *shall seek the regularity properties* of the solution of (8.1) when the form $L(v)$ is defined by data f, g_j, and u_0 of "regularity s", with $-1 < s < 0$. □

13.2 Complements on the Transposition of the Adjoint Isomorphism of Order 1

We again consider (8.1) with $r = 1$; there exists a unique $u \in H^0(Q) = L^2(Q)$ with

$$(13.2) \qquad \langle u, \overline{A^* v - v'} \rangle = L(v) \qquad \forall v \in X^1(Q),$$

where L is a continuous antilinear form on $X^1(Q)$.

We *again* choose L in the form

$$(13.3) \qquad L(v) = \langle f, \bar{v} \rangle + \langle \vec{g}, \overline{\vec{T} v} \rangle + \langle u_0, \overline{v(x, 0)} \rangle.$$

On the other hand the choice of f, $\vec{g} = \{g_0, \ldots, g_{m-1}\}$ and u_0 shall be different from Section 11: *we no longer require* f, g_j *and* u_0 *to be distributions on* Q, Σ *and* Ω. □

Thus we take

$$(13.4) \qquad f \in (H^{2m,1}(Q))'. \quad \square$$

Since, according to Section 2, $v \to \vec{T} v$ is a continuous linear mapping of

$$X^1(Q) \to \prod_{j=0}^{m-1} H^{m_j + 1/2, (m_j + 1/2)/2m}(\Sigma)$$

we may take

$$(13.5)\qquad \vec{g} \in \prod_{j=0}^{m-1} (H^{m_j+1/2,(m_j+1/2)/2m}(\Sigma))'$$

in (13.3) (the spaces $(H^{m_j+1/2,(m_j+1/2)/2m}(\Sigma))'$ *not being* spaces of distributions on Σ). □

Similarly, since $v \to v(x, 0)$ is a continuous linear mapping of $X^1(Q) \to H^m(\Omega)$, we may take

$$(13.6)\qquad u_0 \in (H^m(\Omega))'$$

in (13.3) ($(H^m(\Omega))'$ *not being* a space of distributions on Ω). □

In short:

Theorem 13.1. *Let f, g, u_0 be given with* (13.4), (13.5) *and* (13.6), *under the hypotheses of Section* 1. *There exists a unique $u \in H^0(Q)$ such that* (13.2) *holds with* (13.3). □

13.3 Orientation

We shall seek suitable hypotheses, which depend on s $(-1 < s < 0)$, on f, g and u_0 such that we can show that the solution u of Theorem 13.1 belongs to $H^{2m(s+1),s+1}(Q)$. We shall apply interpolation theory in the following way.

If we consider the mapping $\mathcal{G}$ (Green's operator for problem (13.1))

$$(13.7)\qquad \mathcal{G}: \{f, \vec{g}, u_0\} \to u = \mathcal{G}(f, \vec{g}, u_0),$$

then Section 13.1 and Theorem 13.1 tell us that it is a continuous mapping of

$$(13.8)\qquad \begin{cases} H^0(Q) \times \left\{ \prod_{j=0}^{m-1} H^{2m-(m_j+1/2),1-(m_j+1/2)/2m}(\Sigma) \times H^m(\Omega),\ \mathcal{R.C.} \right\} \\ \to H^{2m,1}(Q) \end{cases}$$

and of

$$(13.9)\qquad (H^{2m,1}(Q))' \times \prod_{j=0}^{m-1} (H^{m_j+1/2,(m_j+1/2)/2m}(\Sigma))' \times (H^m(\Omega))' \to H^0(Q).$$

We recall that

$$\left\{ \prod_{j=0}^{m-1} H^{2m-(m_j+1/2),1-(m_j+1/2)/2m}(\Sigma) \times H^m(\Omega),\ \mathcal{R.C.} \right\}$$

denotes the *subspace* of the product space whose elements satisfy the compatibility relations.

We shall now interpolate between (13.8) *and* (13.9). □

Remark 13.1. The mapping $\mathcal{G}$ is not injective from

$$(H^{2m,1}(Q))' \times \prod_{j=0}^{m-1} (H^{m_j+1/2,(m_j+1/2)/2m}(\Sigma))' \times (H^m(\Omega))' \to H^0(Q),$$

since we can choose $\{f, \vec{g}, u_0\} \neq 0$ with (13.4), (13.5) and (13.6) such that the form $L(v)$ defined in (13.3) $\equiv 0$ on $X^1(Q)$ (since f, $\vec{g}$ and u_0 may have support on Γ). □

14. Some Interpolation Theorems

14.1 Notation. Statement of the Main Result

We set (see (13.8) and (13.9)):

$$A_0 = \left\{\prod_{j=0}^{m-1} H^{2m-(m_j+1/2),1-(m_j+1/2)/2m}(\Sigma) \times H^m(\Omega), \mathcal{R}.\mathcal{C}.\right\} \tag{14.1}$$

$$A_1 = \prod_{j=0}^{m-1} (H^{m_j+1/2,(m_j+1/2)/2m}(\Sigma))' \times (H^m(\Omega))'. \tag{14.2}$$

We also set

$$\mathcal{H}^{\beta}(\Omega) = \begin{cases} H^{\beta}(\Omega) & \text{if } \beta \geqq 0 \\ (H^{-\beta}(\Omega))' & \text{if } \beta \leqq 0 \end{cases} \tag{14.3}$$

$$\mathcal{H}^{\alpha,\alpha/2m}(\Sigma) = \begin{cases} H^{\alpha,\alpha/2m}(\Sigma) & \text{if } \alpha \geqq 0 \\ (H^{-\alpha,-\alpha/2m}(\Sigma))' & \text{if } \alpha \leqq 0. \end{cases} \tag{14.4}$$

Theorem 14.1. *Notation of* (14.1) *to* (14.4). *Let m_j be the order of B_j, $0 \leqq j \leqq m-1$, and assume* (1.6) *to be satisfied. If*

$$\theta > \frac{1}{2} - \frac{1}{4m} - \frac{\min m_j}{2m} = \eta \tag{14.5}$$

then

$$[A_0, A_1]_\theta = \prod_{j=0}^{m-1} \mathcal{H}^{2m(1-\theta)-(m_j+1/2),1-\theta-(m_j+1/2)/2m}(\Sigma) \times \mathcal{H}^{(1-2\theta)m}(\Omega). \tag{14.6}$$

Remark 14.1. In (14.5) we always take $0 < \theta < 1$.

Condition (14.5) is therefore always satisfied if $\eta \leqq 0$, i.e. if $\min_j m_j \geqq m - \frac{1}{2}$, therefore if

$$m_j \geqq m \qquad \forall j. \tag{14.7}$$

Then (14.6) *holds* $\forall \theta \in]0, 1[$. □

Remark 14.2. The *compatibility relations* do not appear in $[A_0, A_1]_\theta$.

In fact these relations would bring in the expressions $B_j u_0|_\Gamma$ for the element $\{g_j, u_0\}$; if $u_0 \in H^{(1-2\theta)m}(\Omega)$, then $B_j u_0|_\Gamma$ has meaning only if the order of $B_j = m_j < (1-2\theta)m - \frac{1}{2}$, therefore if

$$\theta < \frac{1}{2} - \frac{1}{4m} - \frac{m_j}{2m}. \tag{14.8}$$

Thus the compatibility relations *are not defined* if (excluding the ambiguous case of equality)

$$\theta > \frac{1}{2} - \frac{1}{4m} - \frac{m_j}{2m} \quad \forall j,$$

that is if (14.5) is satisfied. □

Remark 14.3. It seems (very likely) that, for *arbitrary* $\theta \in]0, 1[$, $[A_0, A_1]_\theta$ is given by the subspace of

$$\prod_{j=0}^{m-1} \mathscr{H}^{2m(1-\theta)-(m_j+1/2), 1-\theta-(m_j+1/2)2m}(\Sigma) \times \mathscr{H}^{(1-2\theta)m}(\Omega)$$

defined by those compatibility relations which have meaning on this space (with in addition the exceptional cases

$$1 - \theta - \frac{m_j + \frac{1}{2}}{2m} = \frac{1}{2},$$

where integral relations appear).

14.2 Outline of the Proof

The compatibility relations for $\{g_j, u_0\}$ can be expressed by

$$B_j\left(x, 0, \frac{\partial}{\partial x}\right) u_0 = g_j(x, 0), \quad x \in \Gamma. \tag{14.9}$$

Writing $B_j(0) = B_j\left(x, 0, \frac{\partial}{\partial x}\right)$, we introduce

$$V = \{v \mid v \in H^m(\Omega), B_j(0) v = 0 \text{ for all } j \text{ such that } m_j \leqq m-1\}. \ \square \tag{14.10}$$

Remark 14.4. If $m_j \geqq m \ \forall j$, then $V = H^m(\Omega)$. □
Then we introduce

(14.11)

$${}_0H^{2m-(m_j+1/2), 1-(m_j+1/2)/2m}(\Sigma) = \Big\{ v \mid v \in H^{2m-(m_j+1/2), 1-(m_j+1/2)/2m}(\Sigma),$$

$$v(x, 0) = 0 \quad \text{for} \quad x \in \Gamma \quad \text{if} \quad 1 - \frac{m_j + \frac{1}{2}}{2m} > \frac{1}{2} \Big\},$$

(space which *differs* from $H^{\alpha,\beta}_{,0}(\Sigma)$ as defined in Section 2, except if $1 - \frac{m_j + \frac{1}{2}}{2m} < \frac{1}{2}$).

We set

$$(14.12)\qquad \begin{cases} C_0 = \prod_{j=0}^{m-1} {}_0H^{2m-(m_j+1/2),1-(m_j+1/2)/2m}(\Sigma) \times V, \\ D_0 = \prod_{j=0}^{m-1} H^{2m-(m_j+1/2),1-(m_j+1/2)/2m}(\Sigma) \times H^m(\Omega). \end{cases}$$

Remark 14.5. In C_0 the compatibility relations are trivially satisfied, each well-defined term of (14.9) being *zero*. There are no compatibility relations in D_0. □

Evidently $C_0 \subset A_0 \subset D_0$ and hence

$$(14.13)\qquad [C_0, A_1]_\theta \subset [A_0, A_1]_\theta \subset [D_0, A_1]_\theta .$$

In the sequel we shall specify the spaces $[C_0, A_1]_\theta$ and $[D_0, A_1]_\theta$, spaces which coincide when (14.5) holds, from which Theorem 14.1 will follow. □

14.3 First Auxiliary Interpolation Theorem

Theorem 14.2. *We have (in the notation of* (14.4)):

$$(14.14)\qquad [H^{2m-\alpha,1-\alpha/2m}(\Sigma), (H^{\alpha,\alpha/2m}(\Sigma))']_\theta = \mathscr{H}^{(1-\theta)2m-\alpha,1-\theta-\alpha/2m}(\Sigma), \qquad 0 < \alpha < 2m.$$

Proof. (This proof is very similar to the proof of Theorem 12.5 of Chapter 1.)

1) In general, let $\mathscr{V}$ and $\mathscr{H}$ be two Hilbert spaces, $\mathscr{V} \subset \mathscr{H}$, $\mathscr{V}$ dense in $\mathscr{H}$; if $\mathscr{H}$ is identified with its antidual and if $\mathscr{V}'$ is the antidual of $\mathscr{V}$, then we have:

$$\mathscr{V} \subset \mathscr{H} \subset \mathscr{V}'.$$

If we set

$$\mathscr{V}^{1-\eta} = [\mathscr{V}, \mathscr{H}]_\eta,$$

then

$$(14.15)\qquad [\mathscr{V}, \mathscr{V}']_\theta = \begin{cases} \mathscr{V}^{1-2\theta} & \text{if } 1 - 2\theta \geqq 0 \\ (\mathscr{V}^{2\theta-1})' & \text{if } 1 - 2\theta \leqq 0. \end{cases}$$

This is a consequence of the *reiteration theorem* (Chapter 1, Section 6.1), since $[\mathscr{V}, \mathscr{V}']_{1/2} = \mathscr{H}$ (or else can be easily verified directly).

2) Choose an integer μ such that

$$(14.16)\qquad \mu > 2m - \alpha, \qquad \mu > \alpha.$$

By 1) and Section 2.1, we obtain

$$H^{2m-\alpha,1-\alpha/2m}(\Sigma) = [H^{\mu,\mu/2m}(\Sigma), ((H^{\mu,\mu/2m}(\Sigma))']_{\theta_1},$$

where

$$(1 - 2\theta_1)\,\mu = 2m - \alpha$$

and

$$(H^{\alpha,\alpha/2m}(\Sigma))' = [H^{\mu,\mu/2m}(\Sigma), (H^{\mu,\mu/2m}(\Sigma))']_{\theta_2},$$

where

$$(2\theta_2 - 1)\,\mu = \alpha.$$

In order to simplify the notation, set $H^{\mu,\mu/2m}(\Sigma) = E$; then

$$[H^{2m-\alpha,1-\alpha/2m}(\Sigma),\,(H^{\alpha,\alpha/2m}(\Sigma))']_\theta = [[E,E']_{\theta_1},\,[E,E']_{\theta_2}]_\theta = [E,E']_{\theta_3}$$

(by the reiteration theorem), where

$$\theta_3 = (1-\theta)\,\theta_1 + \theta\,\theta_2.$$

Applying (14.15) we find

$$[E,E']_{\theta_3} = \mathscr{H}^{\mu(1-2\theta_3),\mu(1-2\theta_3)/2m}(\Sigma);$$

but

$$(1-2\theta_3)\,\mu = (1-\theta)\,2m - \alpha,$$

from which (14.14) follows. □

14.4 Second Auxiliary Interpolation Theorem

Theorem 14.3. *Let* $0 < \alpha < 2m$. *If*

$$1 - \alpha/2m \neq \tfrac{1}{2}, \quad 1 - \theta - \alpha/2m \neq \tfrac{1}{2},$$

we have:

$$\text{(14.17)}\quad \begin{cases} [H^{2m-\alpha,(2m-\alpha)/2m}_{,0}(\Sigma),\,(H^{\alpha,\alpha/2m}(\Sigma))']_\theta = \\ \quad = H^{(2m-\alpha)(1-\theta)-\alpha\theta,[(2m-\alpha)(1-\theta)-\alpha\theta]/2m}_{,0}(\Sigma) \\ \qquad \textit{if}\ \ (2m-\alpha)(1-\theta) - \alpha\,\theta \geqq 0, \\ \quad = (H^{-[(2m-\alpha)(1-\theta)-\alpha\theta],-[(2m-\alpha)(1-\theta)-\alpha\theta]/2m}(\Sigma))' \\ \qquad \textit{if}\ \ (2m-\alpha)(1-\theta) - \alpha\,\theta \leqq 0. \end{cases}$$

(For the case $1 - \theta - \alpha/2m = \frac{1}{2}$, we have a result of the type given in Theorem 12.6 of Chapter 1.)

Proof. The plan of the proof is analogous to the one for the proof of Theorem 12.6 of Chapter 1.

In order to simplify the notation, we set

$$E_0 = H^{2m-\alpha,(2m-\alpha)/2m}_{,0}(\Sigma),$$
$$E_1 = (H^{\alpha,\alpha/2m}(\Sigma))'.$$

1) For the time being we assume

Lemma 14.1. *Let* μ *be an integer* > 0 *such that* $\mu/2m$ *is an integer. Then*

$$\text{(14.18)}\quad [H^{\mu,\mu/2m}_{,0}(\Sigma),\,(H^{\mu,\mu/2m}(\Sigma))']_{1/2} = H^0(\Sigma).$$

2) Let us now verify that

$$(14.19)\quad \begin{cases} [H^{\mu,\mu/2m}_{,0}(\Sigma), H^0(\Sigma)]_\theta = H^{(1-\theta)\mu,(1-\theta)\mu/2m}_{,0}(\Sigma) \\ \text{if } \dfrac{(1-\theta)\,\mu}{2m} \neq \dfrac{1}{2} + \text{integer}. \end{cases}$$

Indeed

$$(14.20)\quad \begin{cases} H^{\mu,\mu/2m}_{,0}(\Sigma) = H^{\mu/2m}_0(0,T; H^0(\Gamma)) \cap H^0(0,T; H^\mu(\Gamma)) \\ H^0(\Sigma) = H^0(0,T; H^0(\Gamma)) \end{cases}$$

and we may apply the result pertaining to "interpolation of intersections" (Chapter 1, Section 12.1); we obtain

(14.21)

$$[H^{\mu,\mu/2m}_{,0}(\Sigma), H^0(\Sigma)]_\theta = [H^{\mu/2m}_0(0,T; H^0(\Gamma)), H^0(0,T; H^0(\Gamma))]_\theta \cap$$
$$\cap\, [H^0(0,T; H^\mu(\Gamma)), H^0(0,T; H^0(\Gamma))]_\theta.$$

But the first space on the right-hand side of (14.21) is equivalent to (according to Theorem 11.6 of Chapter 1, a result which is valid *without change* for functions taking their values in *the same* Hilbert space)

$$H^{(1-\theta)\mu/2m}_0(0,T; H^0(\Gamma)) \quad \text{if} \quad \frac{(1-\theta)\,\mu}{2m} \neq \text{integer} + \frac{1}{2}.$$

The second space on the right-hand side of (14.21) is equivalent to $H^0(0,T; H^{(1-\theta)\mu}(\Gamma))$ (by (14.17), Chapter 1) and therefore (14.21) shows that

$$[H^{\mu,\mu/2m}_{,0}(\Sigma), H^0(\Sigma)]_\theta = H^{(1-\theta)\mu/2m}_0(0,T; H^0(\Gamma)) \cap H^0(0,T; H^{(1-\theta)\mu}(\Gamma))$$

from which (14.19) follows.

3) *Choose* μ such that $\mu/2m = \text{integer} \geqq 1$. Set:

$$H^{\mu,\mu/2m}_{,0}(\Sigma) = X, \qquad H^{\mu,\mu/2m}(\Sigma) = Y$$

Then, by (14.19), we have:

$$(14.22)\qquad E_0 = [X, H^0(\Sigma)]_{\theta_1}, \qquad (1-\theta_1)\,\mu = 2m - \alpha,$$

and on the other hand

$$(14.23)\qquad H^{\alpha,\alpha/2m}(\Sigma) = [H^{\mu,\mu/2m}(\Sigma), H^0(\Sigma)]_{\theta_2}, \qquad (1-\theta_2)\,\mu = \alpha.$$

Then, according to the duality theorem (Chapter 1, Section 6.2), we have

$$E_1 = [H^0(\Sigma), Y']_{1-\theta_2}$$

and by (14.18):

$$E_1 = [[X, Y']_{1/2}, Y']_{1-\theta_2},$$

that is, by reiteration (Chapter 1, Section 6.1),

$$(14.24)\qquad E_1 = [X, Y']_{1-\theta_2/2}.$$

In the same way, with (14.18), (14.22) yields

$$E_0 = [X, Y']_{\theta_1/2}. \tag{14.25}$$

Then the space $[E_0, E_1]_\theta$ is equivalent to

$$\begin{cases} [X, Y']_{(1-\theta)\theta_1/2+\theta(1-\theta_2/2)} = [X, Y']_\beta, \\ \beta = (1-\theta)\dfrac{\theta_1}{2} + \theta\left(1 - \dfrac{\theta_2}{2}\right), \end{cases} \tag{14.26}$$

according to (14.24), (14.25) and the reiteration result.

If $\beta \leqq \frac{1}{2}$, this is equivalent to

$$[X, [X, Y']_{1/2}]_{2\beta} = [X, H^0(\Sigma)]_{2\beta} = (\text{by (14.19)}) = H^{(1-2\beta)\mu,(1-2\beta)\mu/2m}_{,0}(\Sigma),$$

from which we obtain the first formula (14.17), since

$$(1-2\beta)\,\mu = (1-\theta)\,(2m-\alpha) - \theta\,\alpha.$$

If $\beta > \frac{1}{2}$, this is equivalent to

$$[[X, Y']_{1/2}, Y']_{2\beta-1} = [H^0(\Sigma), Y']_{2\beta-1} = (\text{duality theorem}) = [Y, H^0(\Sigma)]'_{2-2\beta},$$

from which we obtain the second formula (14.17), since

$$(1-(2-2\beta))\,\mu = -[(2m-\alpha)\,(1-\theta) - \alpha\,\theta]. \quad \square$$

Before giving the proof of Lemma 14.1, let us state the following corollary which is fundamental to our present discussion.

Corollary 14.1. *Let* $0 < \alpha < 2m$. *Then if*

$$\frac{2m\,(1-\theta) - \alpha}{2m} < \frac{1}{2},$$

we have (in the notation of (14.11))

$$[{}_0H^{2m-\alpha,(2m-\alpha)/2m}(\Sigma), (H^{\alpha,\alpha/2m}(\Sigma))']_\theta = \mathscr{H}^{2m(1-\theta)-\alpha,(2m(1-\theta)-\alpha)/2m}(\Sigma).$$

Proof. We always have

$$H^{2m-\alpha,(2m-\alpha)/2m}_{,0}(\Sigma) \subset {}_0H^{2m-\alpha,(2m-\alpha)/2m}(\Sigma) \subset H^{2m-\alpha,(2m-\alpha)/2m}(\Sigma)$$

and therefore, if E denotes the desired interpolation space,

$$[H^{2m-\alpha,(2m-\alpha)/2m}_{,0}(\Sigma), (H^{\alpha,\alpha/2m}(\Sigma))']_\theta \subset E \subset$$
$$\subset [H^{2m-\alpha,(2m-\alpha)/2m}(\Sigma), (H^{\alpha,\alpha/2m}(\Sigma))']_\theta.$$

Now the first space is given by (14.47), the second is given (as in Chapter 1, Theorem 12.5, replacing Ω by Σ) by

$$\mathscr{H}^{2m(1-\theta)-\alpha,[2m(1-\theta)-\alpha]/2m}(\Sigma),$$

from which the result follows, since these two spaces are identical for

$$\frac{2m(1-\theta)-\alpha}{2m} < \frac{1}{2}. \quad \square$$

Proof of Lemma 14.1. 1) We always have

$$H^{\mu,\mu/2m}_{,0}(\Sigma) \subset H^{\mu,\mu/2m}(\Sigma),$$

therefore, if we denote the interpolation space on the left-hand side of (14.18) by F, we have:

$$F \subset [H^{\mu,\mu/2m}(\Sigma), (H^{\mu,\mu/2m}(\Sigma))']_{1/2}$$

and this last space is $H^0(\Sigma)$.

Therefore $F \subset H^0(\Sigma)$ and, by duality

$$H^0(\Sigma) \subset [H^{\mu,\mu/2m}(\Sigma), H^{-\mu,-\mu/2m}(\Sigma)]_{1/2}.$$

2) Therefore, we shall have (14.18) if we can show the *inverse inclusion*. For this purpose we apply (compare with Chapter 1, Lemma 12.2)

Lemma 14.2. *There exists an extension operator P with the following properties:*

$$P \in \mathscr{L}(H^{\mu,\mu/2m}(\Sigma); H^{\mu,\mu/2m}(\Sigma_\infty)), \quad \textit{where} \quad \Sigma_\infty = \Gamma \times \mathbf{R}_t,$$

$$P \in \mathscr{L}(H^{-\mu,-\mu/2m}(\Sigma); H^{-\mu,-\mu/2m}(\Sigma_\infty)),$$

$$r_\Sigma P u = u \quad \forall u \in H^{-\mu,-\mu/2m}(\Sigma), \quad \textit{where} \quad r_\Sigma = \textit{operator of restriction to } \Sigma.$$

The existence of this operator P can be shown, by "reflection in t" about $t = 0$ and $t = T$ (the variables corresponding to Γ playing the role of parameters), exactly as in the proof of Lemma 12.2 of Chapter 1. □

Then, by interpolation, we have:

$$P \in \mathscr{L}([H^{\mu,\mu/2m}(\Sigma), H^{-\mu,-\mu/2m}(\Sigma)]_{1/2}; [H^{\mu,\mu/2m}(\Sigma_\infty), H^{-\mu,-\mu/2m}(\Sigma_\infty)]_{1/2}).$$

But

$$[H^{\mu,\mu/2m}(\Sigma_\infty), H^{-\mu,-\mu/2m}(\Sigma_\infty)]_{1/2} = H^0(\Sigma_\infty)$$

and therefore if

$$u \in [H^{\mu,\mu/2m}(\Sigma), H^{-\mu,-\mu/2m}(\Sigma)]_{1/2},$$

we have:

$$r_\Sigma P u = u \in H^0(\Sigma), \quad \text{whence the desired result.} \quad \square$$

14.5 Third Auxiliary Interpolation Theorem

Theorem 14.4. *Let the space* V *be defined by* (14.10) *and let* θ *satisfy* (14.5). *We have*

$$[V, H^m(\Omega)']_\theta = \mathscr{H}^{(1-2\theta)m}(\Omega) \tag{14.27}$$

(where $\mathscr{H}^\beta(\Omega)$ is defined by (14.3)).

Proof. 1) Since $V \subset H^m(\Omega)$, we always have

$$[V, H^m(\Omega)']_\theta \subset [H^m(\Omega), H^m(\Omega)']_\theta = \mathscr{H}^{(1-2\theta)m}(\Omega) \tag{14.28}$$

(Chapter 1, Theorem 12.5).

2) On the other hand

$$H_0^m(\Omega) \subset V,$$

therefore

$$[V, H^m(\Omega)']_\theta \supset [H_0^m(\Omega), H^m(\Omega)']_\theta \tag{14.29}$$

and (Chapter 1, Theorem 12.6 and the end of Remark 12.5)

$$\begin{cases} [H_0^m(\Omega), H^m(\Omega)']_\theta = H_{\bar\Omega}^{(1-2\theta)m}(\mathbf{R}^n) \quad \text{(space of restrictions to} \\ \bar\Omega \text{ of distributions in } H^{(1-2\theta)m}(\mathbf{R}^n) \text{ with support in } \bar\Omega). \end{cases} \tag{14.30}$$

But if

$$-\tfrac{1}{2} < (1-2\theta)\,m < \tfrac{1}{2}, \tag{14.31}$$

this last space *coincides* with $\mathscr{H}^{(1-2\theta)m}(\Omega)$ and (14.29), (14.30) yield the inverse inclusion to (14.28). *Therefore we have already shown the theorem for* θ *satisfying* (14.31).

3) In particular, taking $\theta = \frac{1}{2}$, we have shown that

$$[V, H^m(\Omega)']_{1/2} = H^0(\Omega).$$

For $\theta < \frac{1}{2}$, the reiteration theorem (Chapter 1, Section 6.1) yields

$$[V, H^m(\Omega)']_\theta = [V; [V, H^m(\Omega)']_{1/2}]_{2\theta}$$

and the theorem then follows from the following identity (which will be verified in step 4) below):

$$\begin{cases} [V, H^0(\Omega)]_{\theta_1} = H^{(1-\theta_1)m}(\Omega) \\ \text{if } \theta_1 > 1 - \dfrac{1}{2m} - \dfrac{\min m_j}{m}. \end{cases} \tag{14.32}$$

For $\theta > \frac{1}{2}$, reiteration yields

$$\begin{aligned}[V, H^m(\Omega)']_\theta &= [[V, H^m(\Omega)']_{1/2}, H^m(\Omega)']_{2\theta} \\ &= [H^0(\Omega), H^m(\Omega)']_\theta,\end{aligned}$$

from which the desired result follows upon application of Theorem 12.5 of Chapter 1.

4) *Proof of* (14.32).

Note that the relation $\theta_1 > 1 - \dfrac{1}{2m} - \dfrac{\min m_j}{m}$ may be written

$$m_j > (1 - \theta_1)\, m - \tfrac{1}{2}, \quad \forall j. \tag{14.33}$$

Of course, we have

$$[V, H^0(\Omega)]_{\theta_1} \subset [H^m(\Omega), H^0(\Omega)]_{\theta_1} = H^{(1-\theta_1)m}(\Omega),$$

and therefore we need to show the inverse inclusion. Thus let f be given in $H^{(1-\theta_1)m}(\Omega)$. By Theorem 4.2 of Chapter 1, to show that $f \in [V, H^0(\Omega)]_{\theta_1}$ is equivalent to showing that there exists a function u satisfying

$$\begin{cases} u \in L^2(0, \infty; V) \cap H^s(0, \infty; H^0(\Omega)), \ \dfrac{1}{2s} = \theta_1, \\ u(x, 0) = f(x), \end{cases} \tag{14.34}$$

or, in the notation of Section 2 and setting

$$\begin{cases} B_j(0) = B_j = \dfrac{\partial^{m_j}}{\partial \nu^{m_j}} + \displaystyle\sum_{\beta=0}^{m_j-1} b_{j\beta} \dfrac{\partial^\beta}{\partial \nu^\beta}, \\ b_{j\beta} = \text{tangential differential operator on } \Gamma \\ \qquad \text{(with regular coefficients) of order } \leqq m_j - \beta: \end{cases} \tag{14.35}$$

$$\begin{cases} u \in H^{m,s}(\Omega \times]0, \infty[), \\ B_j u = 0 \quad \text{on} \quad \Gamma \times]0, \infty[, \\ u(x, 0) = f(x). \end{cases} \tag{14.36}$$

Since f is given in $H^{(1-\theta_1)m}(\Omega)$, we can define

$$\frac{\partial^j f}{\partial \nu^j} \in H^{(1-\theta_1)m-j-1/2}(\Gamma) \quad \text{for} \quad j < (1 - \theta_1)\, m - \frac{1}{2}. \tag{14.37}$$

We define g_j to satisfy

$$\begin{cases} g_j \in H^{\mu_j, \nu_j}(\Sigma), \ \dfrac{\mu_j}{m} = \dfrac{\nu_j}{s} = \dfrac{m - j - \frac{1}{2}}{m}, \quad (\Sigma = \Gamma \times]0, \infty[) \\ g_j(x', 0) = \dfrac{\partial^j f}{\partial \nu^j}(x'), \quad x' \in \Gamma, \ \ j < (1 - \theta_1)\, m - \dfrac{1}{2} \end{cases} \tag{14.38}$$

which is possible (in an infinity of ways) according to Chapter 1, Theorem 4.2.

Next we define

(14.39) $$g_j = 0 \quad \text{if} \quad j \geq (1-\theta_1)\, m - \tfrac{1}{2}, \quad j \neq m_j$$

and finally we define g_{m_j} by the system (note that $m_j > (1-\theta_1)\, m - \frac{1}{2}$):

(14.40) $$g_{m_j} + \sum_{\beta=0}^{m_j-1} b_{j\beta}\, g_\beta = 0.$$

Let us verify that

(14.41) $$g_j \in H^{\mu_j, \nu_j}(\Sigma) \qquad \forall j\ (\leq \max m_j).$$

It only remains to be verified for g_{m_j}; we proceed by induction on increasing m_j; we may then assume that $g_\beta \in H^{\mu_\beta, \nu_\beta}(\Sigma)$ in (14.40) and according to Proposition 2.3, $b_{j\beta}\, g_\beta \in H^{\mu_\alpha, \nu_\alpha}(\Sigma)$, where

$$\frac{\mu_\alpha}{\mu_\beta} = \frac{\nu_\alpha}{\nu_\beta} = 1 - \frac{m_j - \beta}{\mu_\beta}, \qquad \text{whence the result.}$$

Thus we have constructed functions g_j satisfying (14.41), (14.40) and (14.38). Then, according to Theorem 2.3 (in the "easy" case in which there are no *global* compatibility relations), there exists a $u \in H^{m,s}(\Omega \times$ $\times\,]0, \infty[)$ with

(14.42) $$\frac{\partial^j u}{\partial \nu^j}(x', t) = g_j(x', t), \qquad x' \in \Gamma,$$

and

$$u(x, 0) = f(x).$$

But (14.42) and (14.40) imply (we have done what is necessary for that): $B_j u = 0$ on $\Gamma \times]0, \infty[$, and therefore we have (14.36). □

Remark 14.5. Result (14.32) can be completed for *all* values of the parameter θ_1 (see Grisvard [8]): *the space* $[V, H^0(\Omega)]_{\theta_1}$ *coincides with the subspace of elements* $u \in H^{(1-\theta_1)m}(\Omega)$ *such that*

(14.43) $$B_j u = 0 \quad \text{for} \quad m_j < (1-\theta_1)\, m - \tfrac{1}{2},$$

(14.44) $$\frac{1}{\sqrt{\varrho}} B_j u \in L^2(\Omega) \quad \text{for} \quad m_j = (1-\theta_1)\, m - \frac{1}{2},$$

(ϱ as in (11.18) of Chapter 1).

The case (14.43) can be shown by a simple modification (in the way of choosing g_j) of the preceeding proof; the case (14.44) can be attributed to the same principle, but with much more delicate technical details (and the use of Theorem 2.2). □

14.6 Proof of Theorem 14.1

In the notation of (14.2) and (14.11), we have:

$$[C_0, A_1]_\theta = \prod_{j=0}^{m-1} [{}_0H^{2m-(m_j+1/2),[2m-(m_j+1/2)]/2m}(\Sigma), \\ (H^{m_j+1/2,(m_j+1/2)/2m}(\Sigma))']_\theta \times [V, H^m(\Omega)']_\theta$$

and applying Corollary 14.1 and Theorem 14.4, this is equivalent to

$$\prod_{j=0}^{m-1} \mathscr{H}^{(1-\theta)2m-(m_j+1/2),1-\theta-(m_j+1/2)/2m}(\Sigma) \times \mathscr{H}^{(1-2\theta)m}(\Omega)$$

if θ satisfies (14.5).

But $[D_0, A_1]_\theta$ is *always* (for any θ) equal to the preceding space, from which the theorem follows by (14.13). □

15. Final Results; Existence of Solutions in the Spaces $H^{2mr,r}(Q)$, $0 < r < 1$. Applications

15.1 Application of the Results of Section 14

We reconsider Section 13.3 with the notation of Section 14.1; then $\mathscr{G}$, defined by (13.7), is a continuous linear mapping of

$$[H^0(Q), (H^{2m,1}(Q))']_\theta \times [A_0, A_1]_\theta \to [H^{2m,1}(Q), H^0(Q)]_\theta.$$

Set $\mathscr{H}^{-2\theta m,-\theta}(Q) = (H^{2m\theta,\theta}(Q))'$.

Applying Theorem 14.1, we obtain:

(15.1) $\left\{\begin{array}{l} \text{for } \theta \in]0,1[\text{ and satisfying (14.5) (see also Remark 14.1),} \\ \mathscr{G} \text{ defined by (13.7) is a continuous linear mapping of} \\ \mathscr{H}^{-2\theta m,-\theta}(Q) \times \prod_{j=0}^{m-1} \mathscr{H}^{2(1-\theta)m-(m_j+1/2),1-\theta-(m_j+1/2)/2m}(\Sigma) \times \\ \qquad \times \mathscr{H}^{(1-2\theta)m}(\Omega) \to H^{(1-\theta)2m,1-}(Q). \end{array}\right.$

It suffices to note that $[H^0(Q), (H^{2m,1}(Q))']_\theta = \mathscr{H}^{-2\theta m,-\theta}(Q)$ (same proof as for Theorem 12.5). □

We can now complete the results of (i) and (ii) of Section 13.1 with the following result:

(iii) $\left\{\begin{array}{l} \textit{if } f \in \mathscr{H}^{2ms,s}(Q),\ g_j \in \mathscr{H}^{2m(s+1)-m_j-1/2,s+1-(m_j+1/2)/2m}(\Sigma), \\ u_0 \in \mathscr{H}^{2(s+1/2)m}(\Sigma), \textit{ then the solution of problem (13.1) belongs} \\ \textit{to } H^{2m(s+1),s+1}(Q); \textit{ and this for} \\ -1 < s < -\max\left[0, \dfrac{1}{2} - \dfrac{1}{4m} - \dfrac{\min m_j}{2m}\right]. \end{array}\right.$

Of course, we must point out that in (iii), u is a solution of problem (13.1) in the sense of Theorem 13.1, i.e. u satisfies

$$\langle u, \overline{A^* v - \bar{v}'}\rangle = \langle f, \bar{v}\rangle + \langle \vec{g}, \overline{\vec{T} v}\rangle_\Sigma + \langle u_0, \overline{v(x, 0)}\rangle_\Omega \qquad \forall v \in X^1(Q).$$

If

$$\frac{1}{2} - \frac{1}{4m} - \frac{\min m_j}{2m} \leqq 0,$$

that is if

$$m_j \geqq m \qquad \forall j, \tag{15.2}$$

then we have obtained *all possible orders of regularity* for $-1 < s < 0$.

If, on the other hand, $\frac{1}{2} - \frac{1}{4m} - \frac{\min m_j}{2m} > 0$, then, in the terminology of Section 13.1, we do not have the optimal result for the order of regularity s, with

$$-\left[\frac{1}{2} - \frac{1}{4m} - \frac{\min m_j}{2m}\right] < s < 0. \tag{15.3}$$

For the case (15.3), *there are compatibility relations* between the data: we shall not specify them here. □

Remark 15.1. We can also give another application of the interpolation results of Section 14; a "parabolic" adaptation of Remark 7.3 of Chapter 2.

We introduce the following spaces:

$$\begin{cases} H^{2m,1}_{B,0}(Q) = \{v \mid v \in H^{2m,1}(Q), B_j v = 0 \text{ on } \Sigma, \\ 0 \leqq j \leqq m-1, v(x,0) = 0\} \\ H^{2m,1}_{C,T}(Q) = \{v \mid v \in H^{2m,1}(Q), C_j v = 0 \text{ on } \Sigma, \\ 0 \leqq j \leqq m-1, v(x,T) = 0\}. \end{cases} \tag{15.4}$$

Then

$$\begin{cases} \frac{\partial}{\partial t} + A \left(\text{resp. } -\frac{\partial}{\partial t} + A^*\right) \text{ is an isomorphism of } H^{2m,1}_{B,0}(Q) \\ (\text{resp. } H^{2m,1}_{C,T}(Q)) \text{ onto } L^2(Q). \end{cases} \tag{15.5}$$

By transposition

$$\left(-\frac{\partial}{\partial t} + A^*\right)^* \text{ is an isomorphism of } L^2(Q) \text{ onto } (H^{2m,1}_{B,0}(Q))' \tag{15.6}$$

and with the aid of Green's formula we easily verify that $\left(-\dfrac{\partial}{\partial t} + A^*\right)^*$ is an *extension* of $\dfrac{\partial}{\partial t} + A$, which we shall still denote by $\dfrac{\partial}{\partial t} + A$. Then, interpolating between (15.5) and (15.6), we obtain:

$$(15.7)\qquad \begin{cases} \dfrac{\partial}{\partial t} + A \text{ is an isomorphism of } [H^{2m,1}_{B,0}(Q), L^2(Q)]_\theta \\ \text{onto } [L^2(Q), (H^{2m,1}_{C,T}(Q))']_\theta . \end{cases}$$

The second space in (15.7) coincides (according to Chapter 1, Section 6.2) with

$$([H^{2m,1}_{C,T}(Q), L^2(Q)]_{1-\theta})'.$$

By variants of the preceding methods, it can be shown that

$$(15.8)\qquad [H^{2m,1}_{C,T}(Q), L^2(Q)]_{1-\theta} = H^{2m\theta,\theta}(Q),$$

if *in particular* $2m\,\theta < \frac{1}{2}$ (in which case there are no more compatibility relations). In this case the space $\mathscr{D}(Q)$ is dense in $H^{2m\theta,\theta}(Q)$ and we have (see Section 8.3)

$$(15.9)\qquad (H^{2m\theta,\theta}(Q))' = H^{-2m\theta,-\theta}(Q);$$

furthermore, since $2m(1-\theta) > 2m - \frac{1}{2}$, we can verify that

$$(15.10)\qquad \begin{cases} [H^{2m,1}_{B,0}(Q), L^2(Q)]_\theta = H^{2m(1-\theta),1-\theta}_{B,0}(Q) \\ = \{v \mid v \in H^{2m(1-\theta),1-\theta}(Q),\ B_j v = 0 \text{ on } \Sigma, \\ 0 \leqq j \leqq m-1,\ v(x,0) = 0\} \end{cases}$$

Therefore:

Theorem 15.1. *Under the hypotheses of Section* 2.1, *the operator* $\dfrac{\partial}{\partial t} + A$ *is an isomorphism of* $H^{2sm,s}_{B,0}(Q)$ *onto* $H^{-2ms,-s}(Q)$, *for* $0 < s < \dfrac{1}{4m}$. □

15.2 Examples; Generalities

Suppose that we have to solve problem (13.1) with

$$(15.11)\qquad g_j \in L^2(\Sigma),\ j = 0, \ldots, m-1$$

(this situation will come up in an essential way in Chapter 6). Then we have to apply (iii) of Section 15.1. Choose s such that (if possible!)

$$(15.12)\qquad L^2(\Sigma) \subset \mathscr{H}^{2(s+1)m-(m_j+1/2),s+1-(m_j+1/2)/2m}(\Sigma),$$

that is such that

$$1 + s - \frac{m_j + \frac{1}{2}}{2m} \leqq 0$$

with s taken as large as possible. Therefore

(15.13) $$s^{\text{optimum}} = -1 + \frac{1}{4m} + \frac{\min m_j}{2m} \ (<0).$$

We then obtain:

Theorem 15.2. *Assume that the g_j's are given so as to satisfy* (15.12). *If we take:*

(15.14) $$f \in \mathscr{H}^{-2m+1/2+\min m_j,\, -1+1/4m+\min(m_j)/2m}(Q),$$

(15.15) $$u_0 \in \mathscr{H}^{-m+1/2+\min m_j}(\Omega),$$

then the solution u of problem (13.1) (in the sense of Theorem 13.1) *satisfies*

(15.16) $$u \in H^{1/2+\min m_j,\,(1/2+\min m_j)/2m}(Q). \quad \square$$

15.3 Examples (I)

Example 1.
Consider

(15.17) $$\begin{cases} \dfrac{\partial u}{\partial t} - \Delta u = f, \\ u|_\Sigma = g_0, \quad u(x,0) = u_0. \end{cases}$$

Then $m_j = m_0 = 0$, $m = 1$ and therefore: *if* $g_0 \in L^2(\Sigma)$, we may take $f \in \mathscr{H}^{-3/2,-3/4}(Q)$, $u_0 \in \mathscr{H}^{-1/2}(\Omega)$ and so

(15.18) $$u \in H^{1/2,1/4}(Q). \quad \square$$

Example 2.
Now consider

(15.19) $$\begin{cases} \dfrac{\partial u}{\partial t} - \Delta u = f, \\ \left.\dfrac{\partial u}{\partial \nu}\right|_\Sigma = g_0, \quad u(x,0) = u_0. \end{cases}$$

Then $m = 1$, $m_j = m_0 = 1$ and therefore if $g_0 \in L^2(\Sigma)$, we may take $f \in \mathscr{H}^{-1/2,-1/4}(Q)$, $u_0 \in \mathscr{H}^{1/2}(\Omega)$ and so

(15.20) $$u \in H^{3/2,3/4}(Q). \quad \square$$

Example 3.

Consider

$$(15.21)\qquad \begin{cases} \dfrac{\partial u}{\partial t} + \Delta^2 u = f, \\ u|_\Sigma = g_0, \ \dfrac{\partial u}{\partial \nu}\Big|_\Sigma = g_1, \quad u(x, 0) = u_0. \end{cases}$$

Then $m = 2$, $m_0 = 0$, $m_1 = 1$, $\min m_j = 0$. Therefore if $g_0, g_1 \in L^2(\Sigma)$, we may take $f \in \mathscr{H}^{-4+1/2,-1+1/8}(Q)$, $u_0 \in \mathscr{H}^{-3/2}(\Omega)$ and so

$$(15.22)\qquad u \in H^{1/2,1/8}(Q). \quad \square$$

Example 4.

Finally consider

$$(15.23)\qquad \begin{cases} \dfrac{\partial u}{\partial t} + \Delta^2 u = f, \\ \dfrac{\partial u}{\partial \nu}\Big|_\Sigma = g_0, \quad \dfrac{\partial \Delta u}{\partial \nu}\Big|_\Sigma = g_1, \quad u(x, 0) = u_0. \end{cases}$$

Then $m = 2$, $m_0 = 1$, $m_1 = 3$, $\min m_j = 1$. Therefore if $g_i \in L^2(\Sigma)$, we may take $f \in \mathscr{H}^{-5/2,-5/8}(Q)$, $u_0 \in \mathscr{H}^{-1/2}(\Omega)$ and so

$$(15.24)\qquad u \in H^{3/2,3/8}(Q). \quad \square$$

15.4 Examples (II)

Consider the (homogeneous) boundary value problem already studied in Chapter 3, Section 4.7.5 in the setting of variational evolution equations:

$$(15.25)\qquad \frac{\partial u}{\partial t} + \Delta^2 u = f \quad \text{in } Q$$

$$(15.26)\qquad u = 0, \quad \frac{\partial \Delta u}{\partial \nu} = 0 \quad \text{on } \Sigma,$$

$$(15.27)\qquad u(x, 0) = u_0(x), \quad u_0 \text{ given in } \Omega.$$

Example (12.25)–(15.27) *also fits the theory studied in the present chapter.*

It is required to verify (1.7). Thus we consider the operator

$$(15.28)\qquad \Delta_x^2 + e^{i\theta} D_y^4 \quad \text{in} \quad \{x_n > 0 \mid x \in \mathbf{R}^n, y \in \mathbf{R}\},$$

with the "boundary operators"

$$(15.29)\qquad B_0 = \text{identity}, \quad B_1 = \frac{\partial}{\partial x_n} \Delta_x.$$

Set

$$\varrho^2 = \xi_1^2 + \cdots + \xi_{n-1}^2, \quad \varrho > 0;$$

the polynomial $A_0(x, \xi + \tau \xi')$, already considered in the elliptic theory, corresponding to (15.28) (see Chapter 2, Section 1), becomes

$$(\varrho^2 + \tau^2)^2 + e^{i\theta} \eta^4, \quad \theta \in \left[-\frac{\pi}{2}, \frac{\pi}{2} \right] \tag{15.30}$$

and the polynomials corresponding to (15.29) are

$$1, \quad \tau(\varrho^2 + \tau^2). \tag{15.31}$$

For $\eta = 0$, we have seen (Chapter 2, Section 9.4) that the polynomials (15.31) are independent modulus $(\tau - \tau_1^+)(\tau - \tau_2^+)$, where

$$\tau_k^+ = \text{root of (15.30) with imaginary part} > 0, \ k = 1, 2.$$

It remains to investigate the case $\eta \neq 0$. Then $\tau_1^+ \neq \tau_2^+$. We must have that

$$a + b\tau_1^+(\varrho^2 + (\tau_2^+)^2) = 0,$$

$$a + b\tau_2^+(\varrho^2 + (\tau_2^+)^2) = 0,$$

imply that $a = b = 0$, therefore that

$$\tau_2^+(\varrho^2 + (\tau_2^+)^2) - \tau_1^+(\varrho^2 + (\tau_1^+)^2) \neq 0,$$

that is

$$\varrho^2 + (\tau_1^+)^2 + \tau_1^+ \tau_2^+ + (\tau_2^+)^2 \neq 0, \tag{15.32}$$

which is easily verified. □

15.5 Some Complements on the Dirichlet Problem

Note that for each particular problem, we can obtain new results by applying the same types of methods as in Chapters 3 and 4 and choosing the spaces as a function of the boundary conditions as well.

For example, consider the Dirichlet problem. Assume the operator given by (1.2), with (1.3), *to be strongly elliptic* in $\bar{\Omega}$ uniformly in $t \in [0, T]$. Then, according to the results of Chapter 3 and the trace theorems of this chapter, we find that the problem

$$A u + \frac{\partial u}{\partial t} = f, \tag{15.33}$$

$$\frac{\partial^j u}{\partial \nu^j} = g_j, \quad j = 0, \ldots, m - 1, \tag{15.34}$$

$$u(x, 0) = u_0, \tag{15.35}$$

admits a unique solution in the space

$$W = \left\{ v \mid v \in L^2(0, T; H^m(\Omega)), \frac{\partial v}{\partial t} \in L^2(0, T; H^{-m}(\Omega)) \right\} \tag{15.36}$$

for all

$$\begin{cases} f \in L^2(0, T; H^{-m}(\Omega)), \quad g_j \in H^{m-j-1/2,(m-j-1/2)/2m}(\Sigma), \\ u_0 \in L^2(\Omega). \end{cases} \tag{15.37}$$

More precisely

$$\begin{cases} \textit{the mapping } u \to \\ \to \left\{ A\, u + \frac{\partial u}{\partial t}; \frac{\partial^j u}{\partial \nu^j}, j = 0, \ldots, m-1; u(x, 0) \right\} \\ \textit{is an isomorphism of } W \textit{ onto} \\ L^2(0, T; H^{-m}(\Omega)) \times \prod_{j=0}^{m-1} H^{m-j-1/2,(m-j-1/2)/2m}(\Sigma) \times L^2(\Omega). \end{cases} \tag{15.38}$$

We also recall that, according to Theorem 6.1, we have

$$\begin{cases} \textit{the mapping } u \to \\ \to \left\{ A\, u + \frac{\partial u}{\partial t}; \frac{\partial^j u}{\partial \nu^j}, j = 0, \ldots, m-1; u(x, 0) \right\} \\ \textit{is an isomorphism of } H^{2m,1}(Q) \textit{ onto} \\ \left\{ L^2(Q) \times \prod_{j=0}^{m-1} H^{2m-m_j-1/2,(2m-m_j-1/2)/2m}(\Sigma) \times H^m(\Omega); \mathcal{R}.\mathcal{C}. \right\}, \end{cases} \tag{15.39}$$

and that, according to Remark 12.3, we also have

$$\begin{cases} \textit{the mapping } u \to \\ \to \left\{ A\, u + \frac{\partial u}{\partial t}, \frac{\partial^j u}{\partial \nu^j}, j = 0, \ldots, m-1; u(x, 0) \right\} \\ \textit{is an isomorphism of } D_P^0(Q) \textit{ onto} \\ \Xi^{-2m,-1}(Q) \times \prod_{j=0}^{m-1} H^{-j-1/2,-(j+1/2)/2m}(\Sigma) \times H^{-m}(\Omega). \end{cases} \tag{15.40}$$

Of course, we can now interpolate between these three isomorphisms, which leads to new interpolation problems such as, for example, the interpretation of the space

$$[L^2(0, T; H^{-m}(\Omega)), \Xi^{-2m,-1}(Q)]_\theta .$$

16. Comments

Boundary value problems for second order parabolic equations have since long ago led to a great number of works; for these equations, we refer the reader to the books of Friedman [1] and Ladyzenskaya Uraltzeva-Solonnikov [1], to Il'in-Kalashnikov-Oleinik [1] and Solonnikov [2] and to the bibliographies of these works.

The boundary value problems for parabolic operators studied in this chapter (and which do not necessarily fit the variational theory of Chapter 3) have been solved by Agranovich and Vishik [1] in the Hilbert setting of the spaces $H^{2mr,r}(Q)$ with $r \geqq 1$.

The method followed here (Sections 1–6) is, in our belief, somewhat simpler than the one adopted by these authors. We apply, on the one hand, rather general abstract results on the solution of the equations by Laplace transform (Sections 3 and 5.3) (variants of Doetsch [1], Garnir [1], Lions [1] and [13], Chapter 11) and, on the other hand, an idea of Agmon and Nirenberg (see Agmon [6], Agmon-Nirenberg [1]) to prove the "basic inequalities" (Sections 4 and 5) (Agranovich and Vishik, *loc. cit.*, prove the inequalities *directly*, i.e. by reconsidering the "usual" proofs of the elliptic case (one could also follow the method of Chapter 2, but being careful to examine the contribution of terms containing the variable "Laplace transform")).

Section 2 gives trace theorems in the *non-isotropic* Sobolev spaces $H^{r,s}(Q)$. Theorems 2.2 and 2.3 are due to Grisvard [4, 8] (and we have adopted Grisvard's method here). Particular cases were obtained independently by Fujivara [2]. For Theorem 2.1, see also Pagni [1]. In Grisvard [8], one also finds trace theorems in Sobolev spaces of $H^{r,s}$-type, *but constructed on* L^p, $p \neq 2$; for other bibliographical references on these spaces and on *non-isotropic* Sobolev spaces, see the Comments to Chapter 1. For non-isotropic Morrey-type spaces see Barozzi [1], Da Prato [3].

Coming back to the boundary value problem, the Hilbert setting (that is the setting of Sobolev spaces $H^{2mr,r}(Q)$ – see Section 2 – constructed on L^2) is, of course, not the only one:

i) We can replace L^2 with $L^p, p \neq 2, 1 < p < \infty$; the corresponding theory applies, see Grisvard [6–8], and Solonnikov [2–4]; the first of these authors uses techniques of functional analysis (Dunford integrals to respresent the "inverse operators", furthermore taken in interpolation spaces); the second author uses a more classical method, starting from the analogues of the Poisson kernels, see also Sobolevskii [3, 4].

We also call attention to the studies on parabolic potentials of Arnese [1, 2], Jones [1–4], followed by Fabes [1], Fabes-Jodeit [1], Fabes-Rivière [1], Jodeit [1], Krée [1–4].

ii) We can also consider the setting of spaces of differentiable functions in the usual sense constructed on Lip_α; see Arima [1].

Once in possession of a "good" theory of regularity of boundary value problems, homogeneous or non-homogeneous (the later ones thanks to suitable trace theorems, given in Section 2) we go over to the "irregular" cases (indispensable for the applications to optimal control problems among others) by the *general method:* transposition and interpolation.

We start *solely* from L^2-results; the technical difficulties already encountered in this chapter with L^2 indicate how considerable the "technical" difficulties are for the cases L^p, $p \neq 2$. Transposition and interpolation are presented in Sections 7–17, the results of which are published here for the first time. In connection with the results of Sections 10.3 and 10.5, see also Pini [1].

Questions connected to regularity in Gevrey-type spaces and to the study of boundary value problems of parabolic type in spaces of ultradistributions will be taken up in Volume III.

Let us call attention to some further questions pertaining to parabolic boundary value problems which we have not treated in this book:

a) Green's functions of parabolic problems; see Dressel [1], Eidelman [1] and the bibliography of this work and Arima [1].

b) Equations with discontinuous coefficients, Harnack's inequalities etc.; see Aronson [1], Aronson-Serrin [1, 2], Guglielmino [1, 2], Kroujkov-Oleinik [1], Kroujkov [1, 2], Ladyzenskaya-Uraltzeva [2], Ladyzenskaya-Solonnikov-Uraltzeva [1], Moser [1].

c) Behavior of solutions at infinity in the time variable (see Comments to Chapter 3) and in x, when Ω is unbounded; see Barozzi [2, 3], Friedman [2], Juberg [1, 2], Tacklind [1], Tikonov [1, 2].

d) Uniqueness and unique extension problems: Agmon-Nirenberg [1, 2], Ito-Yamabe [1], S. Krein [1], Landis [1], Lions-Malgrange [1], Mizohata [5].

e) Study of boundary value problems using the theory of pseudo-differential operators (see Vishik [5], Vishik-Eskin [3]), which furthermore allows the treatment of problems with mixed conditions on Σ (for particular cases, see also the examples 4.7 of Chapter 3) or of *oblique derivative* problems.

f) Boundary value problems for parabolic *systems*; see the books of Friedman [1] and Eidelman [2], the work of Solonnikov [3] and the bibliography of these publications.

g) Boundary value problems in non-cylindrical open sets; see Kaminin-Maslennikova [1], Lions [5, 6], Vishik-Eskin [3]; a method using the theory of semi-groups (results of Kato and Tanabe) and the theory of interpolation has been developed by Krein-Laptiev [1].

h) Questions pertaining to approximation of the solution by finite-difference methods; there is a vast literature on this subject, but the study of the general case still poses numerous problems (see Problem 17.19); for a possible way of approximating solutions of non-homogeneous problems, see Lions [29].

i) Questions pertaining to "regular" and "irregular" points of the boundary and to "parabolic" capacity for equations of the second order, "isolated" and "removable" singularities of the solution: see Aronson [2, 3], Eidelman [2], Pini [13, 14].

17. Problems

17.1 In the text we have left aside the *complete* interpretation of the spaces $\Xi^{2rm,r}(Q)$, $H^{2rm}\,\Xi^r(\Sigma)$ *for non-integer* r.

This problem is not a fortiori solved for spaces of this type constructed on L^p, $p \neq 2$, instead of L^2 (see also Chapter 1, Problem 18.4).

17.2 More complete study of *traces* of u's belonging to $D_P^{-(r-1)}(Q)$, in particular the problem, noted in Section 12.3, of *separating* the operators $\vec{B}\,u$ and $u(x, 0)$ in σ_u.

17.3 Complete the results of "regularity s" with $-1 < s < 0$, obtained in Sections 13–15; this will pose many interesting problems of interpolation between the various spaces under consideration (see also Section 15.5).

17.4 In this whole chapter the question of "exceptional parameters" is not treated in optimal fashion (as was attempted in Chapter 2); as a matter of fact, it seemed to us that the technical difficulties encountered in the study of all exceptional cases (in particular with the *global compatibility relations*) were out of proportion with the interest of the results one might obtain, at least with the methods used in this chapter. It would be interesting to see if other methods (for example transforming the problems to equations for pseudo-differential operators on the boundary – but here the boundary is $\Omega \times \Sigma \ldots$) allow for a less difficult study of these exceptional parameters.

17.5 We have *only* used L^2-theory; one could also use the regularity theorems in the spaces L^p, $p \neq 2$, $1 < p < \infty$ (see the Comments).

Similarly, one could use the regularity theorems in the spaces $\mathrm{Lip}_\alpha(\Omega)$ (see Arima [1]).

Then the technical details are very complicated and we have not undertaken any write-up on this subject.

17.6 It would be of interest to systematically extend all the preceding theory to *general parabolic operators* (see Agranovich-Vishik [1] for the L^2-case and the works cited in the Comments for L^p, $p \neq 2$). The L^2-case probably does not present any essentially new difficulties in relation to what has been done in this chapter.

17.7 A systematic study of the case in which the open set Ω has a boundary $\Gamma = \bigcup_{i=1}^{r} \Gamma_i$ *with* Γ_i's *of different dimensions* remains to be done.

17.8 Also the study of non-homogeneous problems in *non-cylindrical open sets remains to be done* (see Comments).

17.9 It would be of interest to study non-homogeneous problems for "elliptic-parabolic" operators; see Fichera [3], Oleinik [1, 2], Kohn-Nirenberg [1, 3], Hörmander [11].

17.10 In this text, we have not considered non-homogeneous problems with "periodicity conditions" instead of "initial conditions":

$$A u + u' = f,$$

$$B u = g,$$

$$u(0) = u(T)$$

$$(\text{or } u(0) = u(T) + \chi, \quad \chi \text{ given})$$

(see Chapter 3, Section 6).

17.11 One can consider "coupled" problems (elliptic-parabolic in a different sense than in 17.9) as follows:

let $\Omega = \Omega_1 \cap \Omega_2$, Γ_1 = *common* part of the boundary; find $u_i(x, t)$, $i = 1, 2$, defined in $Q_i = \Omega_i \times]0, T[$ with *for example*

$$-\Delta u_1 = f_1(x, t) \quad \text{in} \quad Q_1,$$

$$-\Delta u_2 + \frac{\partial u_2}{\partial t} = f_2(x, t) \quad \text{in} \quad Q_2,$$

transmission conditions on $\Gamma_1 \times]0, T[$ and the "usual" conditions on the other parts of the boundary (and an initial condition on u_2). As was pointed out to us by O. A. Oleinik, this case fits the general study of elliptic-parabolic equations (see Problem 17.9) *with piecewise continuous coefficients.*

17.12 Same question for the problem

$$-\Delta u = f \quad \text{in} \quad Q,$$

the time appearing only in the boundary conditions:

$$\frac{\partial u}{\partial \nu} + \frac{\partial u}{\partial t} = g \quad \text{on} \quad \Sigma,$$

with $u(x, 0)$ given for $x \in \Gamma$ (see Lions [13]).

Here is a method of solution. We bring the problem back to the case "$f = 0$".

Denote by $\mathscr{B}$ the continuous linear operator, mapping $H^{1/2}(\Gamma) \to H^{-1/2}(\Gamma)$, defined as follows: if h is given in $H^{1/2}(\Gamma)$, denote by ω the solution in $H^1(\Omega)$ of

$$\Delta \omega = 0,$$

$$\omega|_\Gamma = h,$$

and set

$$\mathscr{B} h = \left.\frac{\partial \omega}{\partial \nu}\right|_\Gamma .$$

Then the problem to be solved is *equivalent* to the following one: if u_1 denotes the trace of u on Σ, find u_1, solution of

$$\mathscr{B} u_1 + \frac{d u_1}{dt} = g,$$

$$u_1(0) \text{ given.}$$

It is easy to verify that $\mathscr{B}$ is coercive on $V = H^{1/2}(\Gamma)$ in the sense

$$(\mathscr{B} v, v) + \lambda \|v\|^2_{L^2(\Gamma)} \geqq \beta \|v\|_{H^{1/2}(\Gamma)}, \quad \beta > 0, \quad \lambda > 0, \quad \forall v \in V.$$

Now, the general theory applies, either as in Chapter 3 or as in the present chapter (for we have regularity theorems for the first-order elliptic operator $\mathscr{B}$); see also Friedman-Shinbrot [1].

In particular, this yields the results of Garipov [1].

This method extends to the case in which $-\Delta$ is replaced by a second-order elliptic operator $A(t)$ with time-dependent coefficients.

The method also extends to the problem *of the second order in t* (see Chapter 3):

$$-\Delta u = 0 \quad \text{in} \quad Q,$$

$$\frac{\partial u}{\partial \nu} + \frac{\partial^2 u}{\partial t^2} = g \quad \text{on} \quad \Sigma,$$

$$u(x, 0), \quad u_t(x, 0) \quad \text{given for} \quad x \in \Gamma.$$

For this type of problem, also consult Friedman-Shinbrot [1], Odhnoff [2], Pleijel [1].

17.13 We most probably have "stability of traces" upon passing from the operator $A + \partial/\partial t$ to "*elliptically regularized*" operators:

$$A + \frac{\partial}{\partial t} + (-1)^m \varepsilon \frac{\partial^{2m}}{\partial t^{2m}}. \tag{17.1}$$

The process of elliptic regularization (17.1) is justified when A and B_j correspond to *a variational problem* (see Chapter 2, Section 9 and Chapter 3, Section 4). It is very likely – but not proven, it seems – that the process (17.1) leads to solutions u_ε which converge towards the solution u of the evolution problem as $\varepsilon \to 0$, when the hypotheses on A, B_j are as in Section 1.1.

It would not be without interest to obtain the non-homogeneous boundary value problems for the *evolution* case by passage to the limit ($\varepsilon \to 0$) starting from suitable *elliptic* boundary value problems. (Can one in this way reach the L^p-theory, $p \neq 2$?)

It would be of equal interest to extend elliptic regularization to the general setting of Agranovich-Vishik [1]; probably, we would have to consider

$$P(x, t, D_x, \varepsilon z D_t^{2m/l} + D_t), \quad \varepsilon > 0, \quad \left(z = e^{i\theta}, \theta \in \left[-\frac{\pi}{2}, \frac{\pi}{2}\right]\right),$$

with the boundary operators (in x)

$$B_j(x, t, D_x, \varepsilon z D_t^{2m/l} + D_t).$$

17.14 It is certainly technically very difficult to extend the results of this chapter to operators $A + \partial/\partial t$ considered in the spaces

$$L^p(0, T; L^q(\Omega)), \quad p \neq q.$$

(There are some results in this direction by Da Prato and by Grisvard.)

However, such a study might be useful for nonlinear problems.

17.15 Problems analogous to the ones studied in this chapter for the operators $A + \partial/\partial t$, where A has unbounded coefficients; for example with coefficients in $L^p(Q)$ (some results for these operators are given in Baiocchi [7], Guglielmino [2], Lions-Raviart [2]).

17.16 Non-homogeneous problems with "mixed" conditions on Σ or with oblique derivative conditions on Σ (see Comments).

17.17 Non-homogeneous parabolic problems in "weighted" spaces.

17.18 Non-homogeneous problems for parabolic *systems*.

17.19 Questions pertaining to approximation by finite difference methods for the non-homogeneous problems studied in this chapter are still open in general; see Lions [29] for a partial result.

Chapter 5

Hyperbolic Evolution Operators, of Petrowski and of Schroedinger. Hilbert Theory.

The reading of this chapter does not require the knowledge of Chapter 4. It rests essentially on Chapters 1 and 2 and on the elements of Chapter 3.

We call attention to the fact that transposition and interpolation are used in a slightly different order than in Chapters 3 and 4.

1. Application of the Results of Chapter 3 and General Remarks

1.1 Notation. Hypotheses

We shall study the equations

$$A\left(x, t, \frac{\partial}{\partial x}\right) u + \frac{\partial^2 u}{\partial t^2} = f \quad \text{in} \quad Q, \tag{1.1}$$

or

$$A u + u'' = f, \tag{1.1a}$$

with the boundary conditions

$$B_j u = g_j, \quad 0 \leqq j \leqq m - 1, \quad \text{on } \Omega \tag{1.2}$$

and the initial conditions

$$u(x, 0) = u_0(x) \quad \text{in} \quad \Sigma, \tag{1.3}$$

$$u'(x, 0) = u_1(x) \quad \text{in} \quad \Omega \quad \left(u'(x, 0) = \frac{\partial u}{\partial t}(x, 0)\right) \tag{1.4}$$

(in the notation of Chapter 4, Section 1 for Q, Ω, Σ).

The hypotheses on the operator A and on the boundary operators B_j are the following: A is defined by

$$A \varphi = \sum_{|p|, |q| \leqq m} (-1)^{|p|} D_x^p (a_{pq}(x, t) D_x^q \varphi), \tag{1.5}$$

where the functions a_{pq} are given in $\mathscr{D}(\bar{Q})$.

The operator A is assumed to be *symmetric*:

$$(1.6) \qquad A^* = A \quad (\text{i.e. } a_{pq} = \overline{a_{pq}}).$$

Remark 1.1. In fact, for the sequel, it is sufficient that the *principal* part of A be symmetric. For this reason, we shall keep both notations A and A^*. □

Let $a(t; u, v)$ be the continuous sesquilinear form on $H^m(\Omega)$:

$$(1.7) \qquad a(t; u, v) = \sum_{|p|, |q| \leq m} \int_\Omega a_{pq}(x, t)\, D_x^p u\, \overline{D_x^p v}\, dx.$$

We assume the operators B_j to be defined by:

$$(1.8) \quad \begin{cases} B_j = B_j(x, t; D_x) = \sum_{|h| \leq m_j} b_{jh}(x, t)\, D_x^h, \\ \text{where the functions } b_{jh} \text{ are given in } \mathscr{D}(\bar{\Sigma}),\ 0 \leq m_j < 2m, \text{ and,} \\ \text{for every } t_0 \in [0, T], \text{ the system } \{B_j(x, t_0; D_x)\}_{j=0}^{m-1} \\ \text{is normal on } \Gamma. \end{cases}$$

Furthermore, we assume the boundary conditions (1.2) to correspond (see Chapter 2, Section 9.4) to (1.7) and a closed vector subspace V of $H^m(\Omega)$:

$$H_0^m(\Omega) \subset V \subset H^m(\Omega),$$

the form $a(t; u, v)$ *being V-coercive,* i.e.

$$(1.9) \qquad a(t; v, v) \geqq \alpha \|v\|^2_{H^m(\Omega)}, \quad \alpha > 0, \quad \forall v \in V. \ \square$$

Remark 1.2. It would be sufficient to assume

$$\text{principal part of } \ a(t; v, u) \geqq \alpha \|v\|^2_{H^m(\Omega)} - \lambda \|v\|_{H^0(\Omega)} \qquad \forall v \in V$$

and for suitable λ. We shall assume (1.9) in order to simplify some technical points. □

Green's formula (compare with Chapter 4, Section 1) may be written (assuming all functions to be regular):

$$(1.10) \quad (A\varphi + \varphi'', \psi) - (\varphi, A^*\psi + \psi'') = (\vec{S}\varphi, \vec{C}\psi)_\Sigma - (\vec{B}\varphi, \vec{T}\psi)_\Sigma +$$
$$+ (\varphi'(T), \psi(T))_\Omega - (\varphi'(0), \psi(0))_\Omega - (\varphi(T), \psi'(T))_\Omega + (\varphi(0), \psi'(0))_\Omega. \ \square$$

Variational formulation of the problem.

Assume $g_j = 0$, $0 \leqq j \leqq m - 1$, in (1.2).

Then problem (1.1), ..., (1.4) may be stated as follows (see Chapter 3):

$$(1.11) \qquad a(t; u(t), v) + (u''(t), v)_\Omega = (f(t), v)_\Omega \qquad \forall v \in V,$$

$$(1.12) \qquad u(0) = u_0, \quad u'(0) = u_1.$$

($u(t)$ denotes the function $x \to u(x, t)$; similarly for $f(t)$). □

1.2 Application of the Results of Chapter 3

The application of Theorem 8.2 of Chapter 3 to the present setting yields:

if $f \in L^2(0,T;H^0(\Omega)) = L^2(Q)$, if $u_0 \in V$ and if $u_1 \in H(=L^2(\Omega))$,

then u is (after a possible modification *in* t on a set of measure zero) a continuous mapping of $[0,T] \to V$ and its derivative $\partial u/\partial t$ is a continuous mapping of $[0,T] \to H$.

Our tool is again the *transposition of a regularity theorem*, but *as always in this text*, we have the *choice* of the regularity theorem, and this choice is dictated by considerations of (relative ...) *simplicity of* the final results.

In the present case, we shall consider (see Section 4 for a precise statement) the *adjoint problem*:

$$A^* v + v'' = \varphi \quad \text{in} \quad Q = \Omega \times]0,T[\tag{1.13}$$

$$C_j v = 0 \quad \text{on} \quad \Sigma, \quad 0 \leqq j \leqq m-1, \tag{1.14}$$

$$v(x,T) = 0, \quad v'(x,T) = 0. \tag{1.15}$$

If, by an application of the results of Chapter 3 which we have just recalled, *we take* $\varphi \in L^2(Q)$, then v describes a space – say X – and

$$X \subset C^0([0,T];V) \cap C^1([0,T];H). \tag{1.16}$$

For transposition, we need the space described by $\vec{T} v = \{T_j v\}_{j=0}^{m-1}$ as v describes X. Now, the *result* (1.16) *is not on general* (except if the order of $T_j \leqq m-1, \forall j$) *sufficient to define* $\{T_j v\}$; we could still define $T_j v$ by using the *additional* fact that $A^* v + v'' \in L^2(Q)$, but this would complicate matters unnecessarily. □

The situation is as follows: *either* we must *improve* the regularity theorem (1.16), *or* we must establish and apply *other* regularity theorems.

In the next section we shall show that an improvement of (1.16) is essentially impossible, and thus we shall be led to establish other regularity theorems in the sequel.

1.3 A Counter-Example

Assume the operator A to be *independent* of t. *We show that for* $f \in L^2(Q)$, $u_0 = 0$, $u_1 = 0$, *the solution* u of (1.11)–(1.12) does not in general belong to $L^2(0,T;H^{2m}(\Omega))$ (so that, in the notation of (1.16), X *is not* contained in $L^2(0,T;H^{2m}(\Omega))$ and there is no simple "natural" definition of $T_j v$, $\forall j$, $v \in X$).

We first consider w to be the solution of

$$(1.17)\qquad \begin{cases} a(w(t), v) + (w''(t), v)_\Omega = 0 \\ w(0) = 0, \\ w'(0) = w_1, \end{cases}$$

with

$$(1.18)\qquad w_1 \in [D(A), H]_{3/4}.$$

We shall show that *in general* $w \notin L^2(0, T; H^{2m}(\Omega))$. This will in turn imply the truth of our assertion, for we can construct Φ with

$$(1.19)\qquad \begin{cases} \Phi \in L^2(0, T; H^{2m}(\Omega)), & \Phi'' \in L^2(0, T; H^0(\Omega)), \\ \Phi(0) = 0, & \Phi'(0) = w_1 \end{cases}$$

and $w - \Phi$ then satisfies

$$(1.20)\qquad A(w - \Phi) + (w - \Phi)'' = -(A\Phi + \Phi''),$$

$$(w - \Phi)(0) = (w - \Phi)'(0) = 0$$

and therefore

$$(1.21)\qquad w - \Phi = u \quad \text{if} \quad f = -(A\Phi + \Phi'').$$

Now $f \in L^2(Q)$, and *if we had* $u \in L^2(0, T; H^{2m}(\Omega))$, then we would have $w \in L^2(0, T; H^{2m}(\Omega))$ and we shall show that this is not true.

The simplest way is to use the spectral decomposition of A. Let $\mathfrak{h} = \int^{\oplus} \mathfrak{h}(\lambda)\, d\mu(\lambda)$ be a measurable Hilbert sum (see Dixmier [1] and Chapter 1, Section 2.3) and $\mathcal{U}$ a unitary operator mapping $H = L^2(\Omega)$ onto $\mathfrak{h}$ such that *$\mathcal{U}$ diagonalizes A*:

$$(1.22)\qquad \mathcal{U} A v = \lambda\, \mathcal{U} v, \quad \forall v \in D(A).$$

If we set

$$(1.23)\qquad \hat{w}(\lambda, t) = (\mathcal{U} w(t))(\lambda), \quad \hat{w}_1(\lambda) = (\mathcal{U} w_1)(\lambda),$$

then (1.17) is *equivalent to*

$$(1.24)\qquad \begin{cases} \lambda \hat{w}(\lambda, t) + \dfrac{d^2}{dt^2} \hat{w}(\lambda, t) = 0, \\ \hat{w}(\lambda, 0) = 0, \\ \dfrac{d\hat{w}}{dt}(\lambda, 0) = \hat{w}_1(\lambda); \end{cases}$$

therefore

$$\hat{w}(\lambda, t) = \frac{1}{\sqrt{\lambda}} \sin (t\sqrt{\lambda})\, \hat{w}_1(\lambda). \tag{1.25}$$

Since w_1 satisfies (1.18), we have (Chapter 1, Section 2.1)

$$\lambda^{1/4}\, \hat{w}(\lambda) \in \mathfrak{h}. \tag{1.26}$$

If w did belong to $L^2(0, T; H^{2m}(\Omega))$, then, since w satisfies the boundary conditions, we would have: $w \in L^2(0, T; D(A))$ and therefore

$$\lambda\, \hat{w} \in L^2(0, T; \mathfrak{h}).$$

But

$$\int_0^T \| \lambda\, \hat{w}(t) \|_{\mathfrak{h}}^2\, dt = \int_0^T \| \lambda^{1/2}\, \hat{w}_1(\lambda) \| \sin^2\!\left(t\sqrt{\lambda}\right) dt$$

and this, under just the condition (1.26), is generally infinite; which proves our assertion. □

Remark 1.3. Let X_A be the space described by the solution v of (1.13), (1.14), (1.15) as φ describes $L^2(Q)$. *This space depends on* A (here we see an *essential difference* with the parabolic case). Indeed, let A_0 be fixed and $u_0 \in X_{A_0}$, with

$$u_0 \notin L^2(0, T; H^{2m}(\Omega))$$

(which is possible according to what we have just seen); therefore, for a suitable index of differentiation α, $|\alpha| = m$, we have:

$$D_x^{2\alpha}\, u_0 \notin L^2(Q)$$

(since if $D_x^{2\alpha}\, u_0 \in L^2(Q)$, $\forall \alpha$, $|\alpha| = m$, then $u_0 \in L^2(0, T; H^{2m}(\Omega))$).

So let $A = A_0 + \lambda\, D_x^{2\alpha}(-1)^m$; for sufficiently small real λ, A is still a (self-adjoint) elliptic operator and $u_0 \notin X_A$, since

$$A\, u_0 + u_0'' = (A_0\, u_0 + u_0'') + \lambda(-1)^m\, D_x^{2\alpha} u_0 \notin L^2(Q). \quad □$$

2. A Regularity Theorem (I)

Theorem 2.1. *Assume that the hypotheses of Section* 1.1 *are satisfied. Assume that* f *is given with*

$$\left\{\begin{aligned} & f \in L^2(0, T; H^0(\Omega)) = L^2(Q) \\ & f' = \frac{\partial f}{\partial t} \in L^2(0, T; H^0(\Omega)) \\ & (\text{i.e. } f \in H^{0,1}(\Omega)) \end{aligned}\right. \tag{2.1}$$

Assume that u_0, u_1 *are given with*

$$(2.2)\quad \begin{cases} u_0 \in \text{domain of } A\left(x, 0, \frac{\partial}{\partial x}\right) \text{ (defined by } a(0; u, v) \text{ and } V) \\ \text{(therefore, in particular, } u_0 \in H^{2m}(\Omega)), \\ u_1 \in H^m(\Omega). \end{cases}$$

Then the solution u *of problem* (1.11), (1.12) (*which exists and is unique*) (see Chapter 3, Section 8 and the preceding Section) *satisfies*

$$(2.3)\quad \begin{cases} u \in L^2(0, T; H^{2m}(\Omega)), \\ u'' = \frac{\partial^2 u}{\partial t^2} \in L^2(0, T; H^0(\Omega)), \end{cases}$$

that is

$$(2.3\text{a})\quad u \in H^{2m,2}(Q).$$

Proof. 1) Let us (formally) differentiate (1.11) with respect to t; we obtain:

$$(2.4)\quad a(t; u'(t), v) + (u'''(t), v)_\Omega + a'(t; u(t), v) = (f'(t), v)_\Omega$$

where

$$a'(t; u, v) = \sum_{|p|, |q| \leq m} \int_\Omega \left(\frac{\partial}{\partial t} a_{pq}(x, t)\right) D_x^q u \overline{D_x^p v}\, dx.$$

But (1.1) yields

$$(2.5)\quad u''(0) = f(0) - A\left(x, 0, \frac{\partial}{\partial x}\right) u_0 = u_2 \in H^0(\Omega).$$

We then consider the a priori resolution of the problem: find u satisfying (2.4) $\forall v \in V$, with (1.12) and (2.5) (u_2 given).

We shall establish the a priori estimates which (by, for example, the method of Faedo/Galerkin) will show that

$$(2.6)\quad \begin{cases} \text{there exists a (unique) } u \text{ satisfying (2.4), (2.5) and} \\ u' \in L^2(0, T; V), \quad u'' \in L^2(0, T; H^0(\Omega)). \end{cases}$$

For this purpose, we replace v by $u''(t)$ in (2.4); taking twice the real part of the result thus obtained, we get:

$$\frac{d}{dt}\left[a(t; u'(t), u'(t)) + \|u''(t)\|^2_{H^0(\Omega)}\right] + 2\operatorname{Re} a'(t; u(t), u''(t)) -$$
$$- a'(t; u'(t), u'(t)) = 2\operatorname{Re}(f'(t), u''(t))_\Omega$$

from which, by integration over t:

$$(2.7)\quad \begin{cases} a(t; u'(t), u'(t)) + \| u''(t)^2_{H^0(\Omega)} \| = \\ = a(0; u_1, u_2) + \| u_2 \|^2_{H^0(\Omega)} - 2\,\mathrm{Re} \int\limits_0^t a'(\sigma; u(\sigma), u''(\sigma))\, d\sigma + \\ \quad + \int\limits_0^t a'(\sigma; u'(\sigma), u'(\sigma))\, d\sigma + 2\,\mathrm{Re} \int\limits_0^t (f'(\sigma), u''(\sigma))_\Omega\, d\sigma. \end{cases}$$

But

$$\int\limits_0^t a'(\sigma; u(\sigma), u''(\sigma))\, d\sigma = a'(t; u(t), u'(t)) - a'(0; u_0, u_1) - \\ - \int\limits_0^t a'(\sigma; u', u')\, d\sigma - \int\limits_0^t a''(\sigma; u, u')\, d\sigma$$

and (2.7) becomes:

$$(2.8)\quad \begin{cases} a(t; u'(t), u'(t)) + \| u''(t) \|^2_{H^0(\Omega)} = a(0; u_1, u_1) + \| u_2 \|^2_{H^0(\Omega)} + \\ + 3 \int\limits_0^t a'(\sigma; u', u')\, d\sigma + 2\,\mathrm{Re} \int\limits_0^t a''(\sigma, u, u')\, d\sigma - \\ - 2\,\mathrm{Re}\, a'(t; u(t), u'(t)) + 2\,\mathrm{Re}\, a'(0; u_0, u_1) + \\ + 2\,\mathrm{Re} \int\limits_0^t (f', u'')_\Omega\, d\sigma. \end{cases}$$

Using (1.9), we find (the c's denoting various constants and setting $\| v \| = \| v \|_{H^m(\Omega)}$, $|v| = \| v \|_{H^0(\Omega)}$):

$$(2.9)\quad \begin{cases} \| u'(t) \|^2 + |u''(t)|^2 \leqq c \Big[\| u_1 \|^2 + |u_2|^2 + \| u_0 \|\, \| u_1 \| + \\ + \| u(t) \|\, \| u'(t) \| + \int\limits_0^t (\| u' \|^2 + \| u \|\, \| u' \|)\, d\sigma + \\ + \int\limits_0^t |f'|\, |u''|\, d\sigma \Big]. \end{cases}$$

But $u(t) = u_0 + \int\limits_0^t u'(\sigma)\, d\sigma$ yields

$$\| u(t) \| \leqq \| u_0 \| + \int\limits_0^t \| u'(\sigma) \|\, d\sigma$$

and so we easily deduce from (2.9) that

$$\|u'(t)\|^2 + |u''(t)|^2 \leqq c\left[\|u_0\|^2_{H^{2m}(\Omega)} + \|u_1\|^2 + \int_0^t |f'(\sigma)|^2\,d\sigma + \right.$$
$$\left. + \int_0^t (\|u'(\sigma)\|^2 + |u''(\sigma)|^2)\,d\sigma\right]$$

from which we obtain the desired inequality by an application of Gronwall's lemma:

$$\text{(2.10)}\quad \begin{cases} \|u'(t)\|^2 + |u''(t)|^2 \leqq c\left[\|u_0\|^2_{H^{2m}(\Omega)} + \|u_1\|^2 + \int_0^T |f'(\sigma)|^2\,d\sigma\right] \\ 0 \leqq t \leqq T. \end{cases}$$

(2.6) follows, as in Chapter 3, or by the method of Faedo-Galerkin (see Chapter 3, Section 8.2).

2) Next, we note that (2.4) can be written (we have done exactly what was necessary for this):

$$\frac{d}{dt}[a(t; u(t), v) + (u''(t), v)_\Omega - (f(t), v)]_\Omega = 0$$

and since u_2 has been chosen such that

$$a(0, u_0, v) + (u_2, v)_\Omega - (f(0), v) = 0,$$

we see that u satisfies (1.11).

But then

$$\text{(2.11)}\qquad a(t; u(t), v) = (\psi(t), v)_\Omega, \quad \forall v \in V,$$

where

$$\text{(2.12)}\qquad \psi = f - u'' \in L^2(0, T; H^0(\Omega)) = L^2(Q).$$

Let us, for the moment assume

Lemma 2.1. *Under the hypothesis:* $a_{pq} \in \mathcal{D}(\bar{Q})$ *and if* (1.9) *holds* (hypothesis (1.6) is not necessary here), *if* u *satisfies* (2.11) *with* (2.12), *then*

$$\text{(2.13)}\qquad u \in L^2(0, T; H^{2m}(\Omega)).$$

Then (2.13) together with (2.6) yields the desired result. □

Proof of Lemma 2.1. Let $A^{-1}(t)f$ be the solution of

$$a(t; A^{1-}(t)f, v) = (f, v)_\Omega, \quad \forall v \in V, \quad f \text{ given in } L^2(\Omega).$$

We know that under these conditions (see Chapter 2)

$$A^{-1}(t) f \in H^{2m}(\Omega).$$

and according to Remark 4.1 of Chapter 2, $\| A^{-1}(t) f \|_{H^{2m}(\Omega)}$ is bounded in t on every compact set. Therefore, according to the formula

$$A^{-1}(t+h) - A^{-1}(t) = -A^{-1}(t+h)(A(t+h) - A(t))^{-1} A(t),$$

the function $t \to A^{-1}(t) f$ is a continuous (and even C^1) mapping of $[0, T] \to H^{2m}(\Omega)$, $\forall f \in L^2(\Omega)$.

Then the function u, which is given by

$$t \to u(t) = A^{-1}(t) \psi(t),$$

is *measurable* from $[0, T] \to H^{2m}(\Omega)$ and

$$\| u(t) \|_{H^{2m}(\Omega)} \leqq C \| \psi(t) \|_{L^2(\Omega)};$$

therefore, we have (2.13). □

3. Regular Non-Homogeneous Problems

3.1 Statement of the Problem

We now consider problem (1.1), ..., (1.4) with $g_j \neq 0$; Theorem 2.1 yielding *one* solution for the case "$g_j = 0$". We reduce the problem to Theorem 2.1 by the usual method:

assume u_0, u_1, g_j $(0 \leqq j \leqq m-1)$ to be given such that there exists a function w satisfying

$$\text{(3.1)} \quad \begin{cases} \begin{cases} A w + w'' = \varphi \in L^2(0, T; H^0(\Omega)) \\ \text{and} \quad \varphi' \in L^2(0, T; H^0(\Omega)), \end{cases} \\ \text{and the conditions} \\ \begin{cases} B_j w = g_j, \quad 0 \leqq j \leqq m-1, \\ w(0) = u_0, \quad w'(0) = u_1. \end{cases} \end{cases}$$

Then setting

$$\text{(3.2)} \quad u - w = \Phi,$$

problem (1.1), ..., (1.4) is equivalent to

$$\text{(3.3)} \quad \begin{cases} A\Phi + \Phi'' = f - \varphi, \\ B_j \Phi = 0, \quad 0 \leqq j \leqq m-1, \\ \Phi(0) = 0, \quad \Phi'(0) = 0, \end{cases}$$

and we apply Theorem 2.1 to problem (3.3). □

3.2 The Compatibility Relations

We now have to specify (3.1). In order to realize conditions "$\varphi, \varphi' \in L^2(0, T; H^0(\Omega))$" in a "practical" way, we shall require w to satisfy

(3.4) $w \in L^2(0,T;H^{2m}(\Omega))$, $w' \in L^2(0,T;H^{2m}(\Omega))$, $w''' \in L^2(0,T;H^0(\Omega))$.

Proposition 3.1. *The necessary and sufficient condition for w to satisfy* (3.4) *is that*

$$\left\{\begin{array}{l} w(t) = w_0 + \int_0^t \psi(\sigma)\, d\sigma, \\ \psi \in H^{2m,2}(Q), \quad w_0 \in H^{2m}(\Omega). \end{array}\right. \tag{3.5}$$

Proof. If $\psi \in H^{2m,2}(Q)$ (i.e. $\psi \in L^2(0,T;H^{2m}(\Omega))$, $\psi'' \in L^2(0,T;H^0(\Omega))$), then w, given by (3.5), satisfies (3.4).

Conversely, if w satisfies (3.4), then $w' = \psi \in H^{2m,2}(Q)$ and w is a continuous mapping of $[0, T] \to H^{2m}(\Omega)$; therefore $w(0) = w_0 \in H^{2m}(\Omega)$ and we have (3.5). □

Proposition 3.2. *If w satisfies* (3.4), *then we have:*

(3.6)

$$B_j w = g_j \in H^1(0,T;H^{2m-m_j-1/2}(\Gamma)) \cap H^{[3m-(m_j+1/2)]/m}(0,T;H^0(\Gamma)),$$

$$w(x, 0) = w_0 \in H^{2m}(\Omega), \tag{3.7}$$

$$w'(x, 0) = w_1 \in H^{3m/2}(\Omega), \tag{3.8}$$

and the compatibility relations

$$\left\{\begin{array}{l} B_j\left(x, 0, \dfrac{\partial}{\partial x}\right) w_0 = g_j(x, 0), \quad x \in \Gamma, \\ \textit{for all } j \textit{ such that } m_j \neq \dfrac{m-1}{2}, \end{array}\right. \tag{3.9}$$

$$\left\{\begin{array}{l} B_j\left(x, 0, \dfrac{\partial}{\partial x}\right) w_1 + \left(\dfrac{\partial}{\partial t} B_j\left(x, 0, \dfrac{\partial}{\partial x}\right)\right) w_0 \\ = \dfrac{\partial}{\partial t} g_j(x, 0), \ x \in \Gamma, \\ \textit{for all } j \textit{ such that } m_j \leqq \dfrac{3m-1}{2}, \end{array}\right. \tag{3.10}$$

to which global relations (see below) *must eventually be added when m is odd.*

Proof. By (3.5), we obtain

$$B_j w(t) = B_j w_0 + B_j\left(\int_0^t \psi(\sigma)\, d\sigma\right). \tag{3.11}$$

Since $w_0 \in H^{2m}(\Omega)$, we have (Chapter 1, Section 9):

$$B_j w \in H^{2m-m_j-1/2}(\Gamma)$$

and

$$t \to B_j\left(x, t\frac{\partial}{\partial x}\right) w$$

belongs to the space defined in (3.6).

On the other hand

$$B_j\left(\int_0^t \psi(\sigma)\, d\sigma\right) = \sum_{|k| \leq m_j} b_{jh}(x,t) \int_0^t D_x^h \psi(x,\sigma)\, d\sigma \quad (x \in \Gamma).$$

But $\psi \in H^{2m,2}(Q)$ and therefore, according to Theorem 2.1 of Chapter 4, we have

$$\frac{\partial^j \psi}{\partial \nu^j} \in H^{2m-j-1/2,(2m-j-1/2)/m}(\Sigma) =$$

$$= H^0(0,T; H^{2m-j-1/2}(\Gamma)) \cap H^{(2m-j-1/2)/m}(0,T; H^0(\Gamma)),$$

so that

$$\int_0^t \gamma_j \psi(\sigma)\, d\sigma \in H^1(0,T; H^{2m-j-1/2}(\Gamma)) \cap H^{1+(2m-j-1/2)/m}(0,T; H^0(\Gamma)),$$

from which (3.6) follows.

We have already seen (3.7); (3.8) follows from trace theorems for ψ. It remains to show the compatibility relations. For this purpose, we apply Theorem 2.3 of Chapter 4. We obtain:

$$\frac{\partial^k \psi}{\partial \nu^k}\Big|_{t=0} = \gamma_k(\psi(0)), \quad 0 \leq k \leq 2m-1, \quad k \neq \frac{m-1}{2}, \tag{3.12}$$

and

$$\frac{\partial}{\partial t}\frac{\partial^k \psi}{\partial \nu^k}\Big|_{t=0} = \gamma_k(\psi('0)), \quad 0 \leq k < \frac{3m-1}{2}, \tag{3.13}$$

(relations which imply (3.9) and (3.10)), to which we must add the *global relations* (assuming, after local maps, that $\Omega = \{x \mid x_n > 0\}$):

$$\int_0^\infty \int_{R^{n-1}} \left| \frac{\partial^k \psi}{\partial x_n^k}(x', \sigma^2, 0) - \frac{\partial^k \psi}{\partial x_n^k}(x', 0, \sigma^{2m}) \right|^2 dx' \frac{d\sigma}{\sigma} < \infty \tag{3.14}$$

$$\text{if } k = \frac{m-1}{2}$$

and

(3.15)

$$\int_0^\infty \int_{R^{n-1}} \left| \frac{\partial^k}{\partial x_n^k} \psi'(x', \sigma^2, 0) - \frac{\partial}{\partial t} \frac{\partial^k}{\partial x_n^k} \psi(x', 0, \sigma^{2m}) \right|^2 dx' \frac{d\sigma}{\sigma} < \infty$$

$$\text{if} \quad k = \frac{3m-1}{2},$$

relations which *imply* global relations for $B_j w_0$, $B_j w_1$, g_j, if there exists m_j such that

$$m_j = \frac{m-1}{2} \quad \text{or} \quad m_j = \frac{3m-1}{2}. \quad \square$$

Conversely, we have:

Proposition 3.3. *Let* g_j, w_0, w_1 *be given and satisfy* (3.6), (3.7), (3.8), (3.9), (3.10) *and the eventual global relations of Proposition* 3.2. *Then there exists a* w *satisfying* (3.4) *and*

$$B_j w = g_j, \quad w(x, 0) = w_0, \quad w'(x, 0) = w_1.$$

Proof. 1) We first consider $\theta \in H^{2m,2}(\Omega)$ with

$$B_j \theta = g_j' - B_j'\left(x, t, \frac{\partial}{\partial x}\right) w_0,$$

$$\theta(x, 0) = w_1.$$

Such a function exists, according to the compatibility relations and Chapter 4 (Section 2 and part 4 of the Proof of Theorem 14.4).

We shall seek w in the form

$$w = F + w_0 + \int_0^t \theta(\sigma)\, d\sigma,$$

with

$$F \in L^2(0, T; H^{2m}(\Omega)), \quad F' \in H^{2m,2}(Q),$$

$$F(x, 0) = 0, \quad F'(x, 0) = 0,$$

$$B_j F = -\int_0^t \left(B_j'\left(x, \sigma, \frac{\partial}{\partial x}\right) \int_0^\sigma \theta(\sigma_1)\, d\sigma_1 \right) d\sigma = h_j.$$

Then we shall have:

$$w(x, 0) = w_0, \quad w'(x, 0) = \theta(x, 0) = w_1$$

and

$$(B_j w)' = (B_j F)' + B_j \theta + B_j' \left(\int_0^t \theta \, d\sigma \right) + B_j' w_0 = g_j',$$

therefore

$$B_j w - g_j = B_j(0) w_0 - g_j(0) = 0.$$

2) *Construction of F*. Let us order the B_j's in increasing order of m_j. Thus, we replace the conditions "$B_j F = h_j$" with:

$$\frac{\partial^j F}{\partial \nu^j} = 0 \quad \text{if} \quad j \leqq m_0 - 1,$$

$$\frac{\partial^{m_0} F}{\partial \nu^{m_0}} = h_0;$$

next, if $B_1 = \dfrac{\partial^{m_1}}{\partial \nu^{m_1}} + \sum_{\beta=0}^{m_1-1} b_{1\beta} \dfrac{\partial^\beta}{\partial \nu^\beta}$, $b_{1\beta}$ = differential operator of order $\leqq m_1 - \beta$ on Σ, we take

$$\frac{\partial^{m_1} F}{\partial \nu^{m_1}} = h_1 - b_{1m_0} h_0$$

and so on. Therefore

$$\frac{\partial^j F}{\partial \nu^j} = k_j, \qquad 0 \leqq j \leqq 2m - 1,$$

where the k_j's belong to the suitable spaces on Σ and the compatibility relations hold:

$$k_j |_{t=0} = 0, \qquad k_j' |_{t=0} = 0.$$

In order to construct F, we start with the construction of $G \in H^{2m,2}(Q)$ with

$$G(x, 0) = 0, \qquad \frac{\partial^j G}{\partial \nu^j} = k_j'$$

(which is possible by Theorem 2.3 of Chapter 4), and then we take

$$F = \int_0^t G(\sigma) \, d\sigma. \quad \square$$

From the preceding results we deduce

Theorem 3.1. *Assume* (1.6) *and* (1.9) *to be satisfied. Let* f, g_j, u_0, u_1 *be given with*

(3.16) $$f \in H^{0,1}(Q),$$

(3.17) $$g_j \in H^1(0, T; H^{2m-m_j-1/2}(\Gamma)) \cap H^{[3m-(m_j+1/2)]/m}(0, T; H^0(\Gamma)),$$

(3.18) $$u_0 \in H^{2m}(\Omega),$$

(3.19) $$u_1 \in H^{3m/2}(\Omega),$$

and the compatibility relations of Proposition 3.2. *Then the solution* u *of problem* (1.1), ..., (1.4) *satisfies*

$$u \in H^{2m,2}(Q). \quad \square \tag{3.20}$$

Remark 3.1. We can verify that:

$$\begin{cases} H^{3m-(m_j+1/2),[3m-(m_j+1/2)]/m}(\Sigma) \subset \\ \subset H^1(0, T; H^{2m-m_j-1/2}(\Gamma)) \cap H^{[3m-(m_j+1/2)]/m}(0, T; H^0(\Gamma)). \end{cases} \tag{3.21}$$

We therefore a fortiori have (3.20) if we replace (3.17) with the condition

$$g_j \in H^{3m-m_j-1/2,[3m-(m_j+1/2)]/m}(\Sigma). \quad \square \tag{3.22}$$

3.3 The Case of the Dirichlet Problem

We consider the *Dirichlet problem*, that is

$$A u + u'' = f \quad \text{in} \quad Q, \tag{3.23}$$

$$\frac{\partial^j u}{\partial \nu^j} = g_j, \quad 0 \leqq j \leqq m-1, \quad \text{on} \quad \Sigma, \tag{3.24}$$

$$\begin{cases} u(x, 0) = u_0 \quad \text{in} \quad \Omega \\ \dfrac{\partial u}{\partial t}(x, 0) = u_1 \quad \text{in} \quad \Omega. \end{cases} \tag{3.25}$$

We have

Theorem 3.2. *Assume that* (1.6) *and* (1.9) *hold. Let* f, g_j, u_0, u_1 *be given with*

$$f \in L^2(Q), \tag{3.26}$$

$$g_j \in H^{2m-j-1/2, (2m-j-1/2)/m}(\Sigma), \quad 0 \leqq j \leqq m-1, \tag{3.27}$$

$$\begin{cases} u_0 \in H^{3m/2}(\Omega), \\ u_1 \in H^{m/2}(\Omega), \end{cases} \tag{3.28}$$

with the compatibility relations such that there exists a $w \in H^{2m,2}(Q)$ *satisfying*

$$\frac{\partial^j w}{\partial \nu^j} = g_j, \quad 0 \leqq j \leqq m-1, \quad w(x, 0) = u_0(x), \quad w'(x, 0) = u_1(x)$$

(see Chapter 4, Section 2, Theorem 2.3).

Then the solution of problem (3.23), (3.24), (3.25) *satisfies*

$$u \in H^{m,1}(Q). \tag{3.29}$$

Remark 3.2. Here the compatibility relations are (see Theorem 2.3, Chapter 4):

(3.30) $$g_j|_{t=0} = \gamma_j u_0 \quad \text{for} \quad 0 \leqq j \leqq m-1$$

(3.31) $$\frac{\partial}{\partial t} g_j|_{t=0} = \gamma_j u_1 \quad \text{for} \quad 0 \leqq j \leqq \frac{m-1}{2}$$

and the following global relation when

(3.32) $$j = \frac{m-1}{2}, \quad m \text{ odd:}$$

(3.33) $$\int_0^\infty \int_{\mathbf{R}^{n-1}} \left| \frac{\partial^j u_1}{\partial x_n^j}(x', \sigma^2) - \frac{\partial g_j}{\partial t}(x', \sigma^{2m}) \right|^2 dx' \frac{d\sigma}{\sigma} < \infty,$$

(where we assume that $\Omega = \{x \mid x_n > 0\}$, to which we are led by local maps).

Proof of Theorem 3.2. The proof is immediate; having chosen a function w as in the statement of the theorem, the function $\Phi = u - v$ must satisfy:

$$A\,\Phi + \Phi'' = f - (A\,w + w''),$$

$$\frac{\partial^j \Phi}{\partial \nu^j} = 0, \quad 0 \leqq j \leqq m-1,$$

$$\Phi(x, 0) = 0, \qquad \frac{\partial \Phi}{\partial t}(x, 0) = 0.$$

Since $w \in H^{2m,2}(Q)$, the function $f - (A\,w - w'')$ belongs to $L^2(Q)$ and therefore Theorem 8.2 of Chapter 3 applies. □

4. Transposition

4.1 Adjoint Isomorphism

Consider the "adjoint problem":

(4.1) $$A^* v + v'' = \varphi$$

with

(4.2) $$\varphi \in H^{0,1}_{0,0}(Q) = \text{closure of } \mathscr{D}(Q) \text{ in } H^{0,1}(Q),$$

and the homogeneous conditions

(4.3) $$C_j v = 0, \quad 0 \leqq j \leqq m-1,$$

(4.4) $$v(T) = 0, \quad v'(T) = 0. \;\square$$

Remark 4.1. For φ, element of $H^{0,1}(Q)$, to belong to $H^{0,1}_{0,0}(Q)$, it is necessary and sufficient that

$$\varphi(x,0) = \varphi(x,T) = 0. \quad \square$$

If we *define*

$$(4.5)\quad \begin{cases} X = \text{space described by } v, \text{ solution of } (4.1),\ldots,(4.4), \text{ as } \varphi \\ \text{describes } H^{0,1}_{0,0}(Q), \text{ provided with the translated topology,} \end{cases}$$

then we see that

$$(4.6)\quad A^* + D_t^2 \text{ is an isomorphism of } X \text{ onto } H^{0,1}_{0,0}(Q).$$

It is this "adjoint isomorphism" which we shall now transpose. □
By Theorem 1.1, we have:

$$(4.7)\quad X \subset H^{2m,2}(Q). \quad \square$$

4.2 Transposition

From (4.6) follows (compare with Section 13 of Chapter 4, for instance):

$$(4.8)\quad \begin{cases} \text{if } v \to L(v) \text{ is a continuous antilinear form on } X, \text{ then there} \\ \text{exists a unique } u \in (H^{0,1}_{0,0}(Q))' \textit{ satisfying} \\ \langle u, \overline{(A^* + D_t^2)\,v}\rangle = L(v), \quad \forall v \in X, \\ \text{where } \langle\ ,\ \rangle \text{ denotes the duality between } (H^{0,1}_{0,0}(Q))' \text{ and} \\ H^{0,1}_{0,0}(Q). \end{cases}$$

If, as we have already done previously (Chapter 4, Section 8.3), we *define*

$$(4.9)\quad H^{0,-1}(Q) = (H^{0,1}_{0,0}(Q))',$$

then we can state (4.8) in the equivalent form:

$$(4.8\,a)\quad \begin{cases} \textit{for every continuous antilinear form } v \to L(v) \textit{ on } X, \textit{ there exists} \\ \textit{a unique } u \in H^{0,-1}(Q) \textit{ satisfying} \\ \langle u, \overline{(A^* + D_t^2)\,v}\rangle = L(v), \quad \forall v \in X. \quad \square \end{cases}$$

Remark 4.2. We recall that every element h of $H^{0,-1}(Q)$ may be represented (non-uniquely) by

$$(4.10)\quad h = h_0 + \frac{\partial}{\partial t} h_1, \quad h_i \in L^2(Q). \quad \square$$

As we have already seen in Chapter 4, the problem now is to *choose* L. □

4.3 Choice of L

Formally at first

$$L(v) = \langle f, \bar{v}\rangle + \langle \vec{g}, \overline{\vec{T} v}\rangle + \langle u_1, \overline{v(0)}\rangle - \langle u_0, \overline{v'(0)}\rangle, \tag{4.11}$$

where the brackets are taken according to suitable dualities.

The choice of f, $\vec{g}$, u_1, u_0 follows from (4.7).

First $v \to \langle f, \bar{v}\rangle$ is evidently, according to (4.7), a *continuous* anti-linear form on X, if we take

$$f \in (H^{2m,2}(Q))'. \quad \square \tag{4.12}$$

Remark 4.3 (Compare with Section 13.2, Chapter 4). Thus f *is not* a distribution on Q. □

Next we note that:

$$\begin{cases} v \to \vec{T} v \quad \text{is a continuous linear mapping of} \\ H^{2m,2}(Q) \to \prod_{j=0}^{m-1} H^{2m-(2m-m_j-1)-1/2,(2m-(2m-m_j-j)-1/2)/m}(\Sigma) \\ = \prod_{j=0}^{m-1} H^{m_j+1/2,(m_j+1/2)/m}(\Sigma), \end{cases} \tag{4.13}$$

and consequently, we may take

$$\vec{g} \in \prod_{j=0}^{m-1} (H^{m_j+1/2,(m_j+1/2)/m}(\Sigma))' \tag{4.14}$$

in (4.11).

(As in Remark 4.3, g *is not* a distribution on Σ.) □

Finally, $v \to v(0)$ (resp. $v \to v'(0)$) is a continuous linear mapping of $H^{2m,2}(Q) \to H^{3m/2}(\Omega)$ (resp. $H^{m/2}(\Omega)$) and therefore, according to (4.7), we may take:

$$u_0 \in (H^{m/2}(\Omega))', \quad u_1 \in (H^{3m/2}(\Omega))' \tag{4.15}$$

in (4.11) (and here again, the u_j's *are not* distributions on Ω).

4.4 Conclusion

The above results may be summarized in

Theorem 4.1. *Assume that* (1.6), (1.8) *and* (1.9) *are satisfied. Let* f, $\vec{g}$, u_0, u_1 *be given with* (4.12), (4.14), (4.15). *There exists a unique* $u \in H^{0,-1}(Q)$ *such that:*

$$\langle u, \overline{(A^* + D_t^2) v}\rangle = \langle f, \bar{v}\rangle + \langle \vec{g}, \overline{\vec{T} v}\rangle + \langle u_1, \overline{v(0)}\rangle - \langle u_0, \overline{v'(0)}\rangle \quad \forall v \in X. \quad \square \tag{4.16}$$

Remark 4.4. In (4.16), the term on the left of the equality sign is taken in the duality between $H^{0,-1}(Q)$ and $H^{0,1}_{0,0}(Q)$. On the right of the equality sign, the first term is taken in the duality between $(H^{2m,1}(Q))'$ and $H^{2m,1}(Q)$, the second between

$$\prod_{j=0}^{m-1} (H^{m_j+1/2,(m_j+1/2)/m}(\Sigma))' \quad \text{and} \quad \prod_{j=0}^{m-1} H^{m_j+1/2,(m_j+1/2)/m}(\Sigma),$$

the third between $(H^{3m/2}(\Omega))'$ and $H^{3m/2}(\Omega)$ and the last between $(H^{m/2}(\Omega))'$ and $H^{m/2}(\Omega)$. □

Remark 4.5. The interpretation of (4.16) in terms of the "usual" boundary value problems is not considered here (see Problem 14.1). Under different hypotheses on f, $\vec{g}$, u_0, u_1, this problem is studied in Section 11. □

Remark 4.6. Taking account of (1.10), the signs in (4.16) are chosen uch that, *formally*, the solution u of (3.16) satisfies (4.1), . . ., (4.4). □

5. Interpolation

5.1 Statement of the Problem

Consider the mapping $\mathscr{G}$[(1)]:

$$\mathscr{G}: f, \vec{g}, u_0, u_1 \to u. \tag{5.1}$$

In order to simplify the notation, set

$$\begin{cases} A_0 = \Big\{\vec{g}, u_0, u_1 \mid \vec{g} \in \prod_{j=0}^{m-1} H^{3m-m_j-1/2,(3m-m_j-1/2)/m}(\Sigma), \\ u_0 \in H^{2m}(\Omega), \quad u_1 \in H^{3m/2}(\Omega), \quad \vec{g}, u_0, u_1 \ \text{ satisfy the } \mathscr{R}.\mathscr{C}.\Big\} \end{cases} \tag{5.2}$$

$$A_1 = \prod_{j=0}^{m-1} (H^{m_j+1/2,(m_j+1/2)/m}(\Sigma))' \times (H^{m/2}(\Omega))' \times (H^{3m/2}(\Omega))', \tag{5.3}$$

where each space is provided with the norm $(\|\vec{g}\|^2 + \|u_0\|^2 + \|u_1\|^2)^{1/2}$, and where each norm for $\vec{g}$, u_0, u_1 is taken in the appropriate space.

Then, according to Theorems 3.1 and 4.1, the linear mapping (5.1) has the following properties:

$$\begin{cases} \mathscr{G} \text{ is a continuous mapping of } H^{0,1}(Q) \times A_0 \to H^{2m,2}(Q), \\ \mathscr{G} \text{ is a continuous mapping of } (H^{2m,2}(Q))' \times A_1 \to H^{0,-1}(Q). \end{cases} \tag{5.4}$$

[(1)] As in Remark 13.1 of Chapter 4, we must be careful to remember that this mapping is not injective.

Therefore by interpolation of (5.4), we have:

$$(5.5)\quad \begin{cases} \mathscr{G} \text{ is a continuous mapping of} \\ [H^{0,1}(Q), (H^{2m,2}(Q))']_\theta \times [A_0, A_1]_\theta \to [H^{2m,2}(Q), H^{0,-1}(Q)]_\theta \end{cases}$$

We need to make (5.5) explicit, and we shall *partially* accomplish this.

5.2 Some Interpolation Results

We start by giving an interpretation of the space $[H^{2m,2}(Q), H^{0,-1}(Q)]_\theta$.

But we must be careful: *it is not true* that this space coincides with the space of restrictions to Q of the v's $\in L^2(\mathbf{R}_t; H^{2m(1-\theta)}(\Omega))$ such that

$$\frac{1}{1+|\tau|}\hat{v} \in L^2(\mathbf{R}_\tau; H^0(\Omega)),$$

$\hat{v}$ = Fourier transform in t of v. The correct interpretation is:

Theorem 5.1. *The following identities hold:*

$$(5.6)\quad \begin{cases} [H^{2m,2}(Q), H^{0,-1}(Q)]_\theta = H^{-\theta}(0,T; H^{2m(1-\theta)}(\Omega)) \cap \\ \cap H^{2-3\theta}(0,T; H^0(\Omega)), \quad \theta \neq \frac{1}{2} \quad \text{and} \quad \theta \neq \frac{5}{6}; \end{cases}$$

$$(5.7)\quad [H^{2m,2}(Q), H^{0,-1}(Q)]_{1/2} = H_{00}^{-1/2}(0,T; H^m(\Omega)) \cap \cap H^{1/2}(0,T; H^0(\Omega))$$

(where $H_{00}^{-1/2}(0,T;X) = (H_{00}^{1/2}(0,T;X))'$ if X is a Hilbert space and $H_{00}^{1/2}(0,T;X) = [H_0^1(0,T;X), H^0(0,T;X)]_{1/2}$; see Chapter 1, Section 11);

$$(5.8)\quad [H^{2m,2}(Q), H^{0,-1}(Q)]_{5/6} = H^{-5/6}(0,T; H^{m/3}(\Omega)) \cap \cap H_{00}^{-1/2}(0,T; H^0(\Omega)).$$

Proof. We first replace the cylinder $Q = \Omega \times]0, T[$ with $Q_\infty = \Omega \times \mathbf{R}_t$; we already assume that $Q = \Omega \times]0, \infty[$ (otherwise make *two* extensions by symmetry, instead of one as below, and use the truncations).

Set

$$p\,u(t) = \begin{cases} u(t) & \text{if } t > 0 \\ \alpha_1 u(-t) + \alpha_2 u(-2t) + \alpha_3 u(-3t) & \text{if } t < 0, \end{cases}$$

with

$$\alpha_1 + \alpha_2 + \alpha_3 = 1$$

$$-\alpha_1 - 2\alpha_2 - 3\alpha_3 = 1$$

$$-\alpha_1 - \frac{\alpha_2}{2} - \frac{\alpha_3}{3} = 1.$$

$u \to p\, u$ is a continuous linear mapping of

$$H^{2m,2}(Q) \to H^{2m,2}(Q_\infty)$$

and of

$$H^{0,-1}(Q) = H^{-1}(0,T;H^0(\Omega)) \to H^{0,-1}(Q_\infty) = H^{-1}(\mathbf{R}_t;H^0(\Omega))$$

(if we note that the *transpose of* p maps

$$H^1(\mathbf{R}_t;\, H^0(\Omega)) \to H^1_0(]0,T[;\, H^0(\Omega))).$$

Therefore $u \to p\, u$ is a continuous linear mapping of

$$[H^{2m,2}(Q),\, H^{0,-1}(Q)]_\theta \to [H^{2m,2}(Q_\infty),\, H^{0,-1}(Q_\infty)]_\theta$$

and since the *restriction* of Q_∞ to Q is a continuous linear mapping of

$$H^{2m,2}(Q_\infty) \text{ (resp. } H^{0,-1}(Q_\infty)) \to H^{2m,2}(Q) \text{ (resp. } H^{0,-1}(Q))$$

and satisfies $r\, p\, u = u$, we finally obtain:

$$[H^{2m,2}(Q),\ H^{0,-1}(Q)]_\theta = \text{space of restrictions to } Q \text{ of}$$
$$[H^{2m,2}(Q_\infty),\, H^{0,-1}\ (Q_\infty)]_\theta.$$

But using the extensions of $\Omega \to \mathbf{R}^n$ (see Chapter 1, Sections 8 and 9), this last space coincides with the space of restrictions to Q_∞ of

$$[H^{2m,2}(\mathbf{R}^n_x \times \mathbf{R}_t),\, H^{0,-1}(\mathbf{R}^n_x \times \mathbf{R}_t)]_\theta = X_\theta.$$

2) Now, by an application of Fourier transformation, the characterization of X_θ is immediate. If $\hat{U}$ denotes the Fourier transform in x *and* t of U, we have:

$$U \in X_\theta \Leftrightarrow (1 + |\xi|^{2m} + |\tau|^2)^{1-\theta} \frac{1}{(1+|\tau|)} \hat{U} \in L^2(\mathbf{R}^n_\xi \times \mathbf{R}_\tau),$$

and thus

$$X_\theta = H^{-\theta}(\mathbf{R}_t;\, H^{2m(1-\theta)}(\mathbf{R}^n_x)) \cap H^{2-3\theta}(\mathbf{R}_t;\, H^0(\mathbf{R}^n_x)).$$

Since the space of restrictions to Ω of $H^s(\mathbf{R}^n_x)$, $s > 0$, is $H^s(\Omega)$, it follows that:

$$[H^{2m,2}(Q),\, H^{0,-1}(Q)]_\theta = \text{space of restrictions to }]0,T[\text{ of}$$
$$H^{-\theta}(\mathbf{R}_t;\, H^{2m(1-\theta)}(\Omega)) \cap H^{2-3\theta}(\mathbf{R}_t;\, H^0(\Omega)).$$

3) In order to complete the proof of the theorem, it only remains to verify:
if $\mathscr{K}$ is a Hilbert space, then the space of restrictions to $]0,T[$ of $H^s(\mathbf{R}_t, \mathscr{K})$, $-1 < s < 2$, coincides with $H^s(0,T;\mathscr{K})$ for $s \neq -\frac{1}{2}$, and with

$$H^{-1/2}_{00}(]0,T[;\mathscr{K}) \quad \text{for} \quad s = -\tfrac{1}{2}.$$

For $s \geqq 0$, this is Theorem 9.1 of Chapter 1 (for functions with values in a Hilbert space). It remains to study the case $-1 < s < 0$. First of all, if $U \in H^s(\mathbf{R}_t; \mathscr{K})$ and if $\mathcal{O} =]0, T[$, its restriction $U_{\mathcal{O}}$ to $\mathcal{O}$ is defined by

$$(5.9) \qquad \langle U_{\mathcal{O}}, \varphi\rangle = \langle U, \tilde{\varphi}\rangle, \quad \varphi \in \mathscr{D}(\mathcal{O}),$$

$$\tilde{\varphi} = \text{extension of } \varphi \text{ by } 0 \text{ outside } \mathcal{O}.$$

Since $\varphi \to \tilde{\varphi}$ is a continuous mapping of

$$H_0^{-s}(\mathcal{O}) \to H^{-s}(\mathbf{R}), \quad s \neq -\tfrac{1}{2},$$

$$H_{00}^{1/2}(\mathcal{O}) \to H^{1/2}(\mathbf{R}),$$

it follows that

$$U_{\mathcal{O}} \in H^s(0, T; \mathscr{K}), \quad s \neq -\tfrac{1}{2}$$

$$U_{\mathcal{O}} \in H_{00}^{-1/2}(]0, T[; \mathscr{K}) \text{ if } s = -\tfrac{1}{2}.$$

We still have to show that the mapping $U \to U_{\mathcal{O}}$ is *surjective*. We know (Chapter 1, Section 12.2)[1] that there exists a continuous linear mapping

$$u \to P u$$

of

$$H^0(0, T; \mathscr{K}) \to H^0(\mathbf{R}_t; \mathscr{K}) \quad \text{and of} \quad H^{-1}(0, T; \mathscr{K}) \to H^{-1}(\mathbf{R}_t; \mathscr{K})$$

such that $(P u)_{\mathcal{O}} = u$. By interpolation, P is a continuous linear mapping of

$$[H^0(0, T; \mathscr{K}), H^{-1}(0, T; \mathscr{K})]_\theta \to H^{-\theta}(\mathbf{R}_t; \mathscr{K})$$

and the first space is (Chapter 1, Section 12)

$$H^{-\theta}(0, T; \mathscr{K}), \quad \theta \neq \tfrac{1}{2},$$

$$H_{00}^{-1/2}(0, T; \mathscr{K}), \quad \text{if } \theta = \tfrac{1}{2}.$$

Whence the desired result. □

We have not solved the problem of the complete characterization (different from the definition!) of $[A_0, A_1]_\theta$; see Problem 14.1. Here, we give *a particular case* (which yields interesting applications to optimal control theory, Chapter 6, Section 11). We consider the case:

$$u_0 = 0, \quad u_1 = 0.$$

We introduce

$$(5.10) \quad \begin{cases} E_j(\Sigma) = \text{space described by } g_j \text{ when } \{\vec{g}, 0, 0\} \text{ belongs to } A_0, \\ \text{with } \vec{g} = \{0, \ldots, 0, g_j, 0, \ldots, 0\}, \text{ provided with the topo-} \\ \text{logy induced by } A_0; \end{cases}$$

[1] Lemma 12.2 of Chapter 1 is valid for vector-valued functions.

and

$$A_{00} = \prod_{j=0}^{m-1} E_j(\Sigma). \tag{5.11}$$

Next, we set

$$A_{10} = \prod_{j=0}^{m-1} (H^{m_j+1/2,(m_j+1/2)/m}(\Sigma))'. \tag{5.12}$$

Then, we have

$$[A_{00}, A_{10}]_\theta = \prod_{j=0}^{m-1} [E_j(\Sigma), (H^{m_j+1/2,(m_j+1/2)/m}(\Sigma))']_\theta. \tag{5.13}$$

Replace u_0, u_1 with 0 in (5.1); thus, we consider

$$f, \vec{g}, 0, 0 \xrightarrow{\pi_0} u, \tag{5.14}$$

which is a continuous linear mapping of (we do not write the multiplication by $\{0\} \times \{0\}$)

$$H^{0,1}(Q) \times A_{00} \to H^{2m,2}(Q)$$

and

$$(H^{2m,2}(Q))' \times A_{10} \to H^{0,-1}(Q),$$

therefore of

$$[H^{0,1}(Q), (H^{2m,2}(Q))']_\theta \times [A_{00}, A_{10}]_\theta \to [H^{2m,2}(Q), H^{0,-1}(Q)]_\theta, \tag{5.15}$$

where the last space is characterized by Theorem 5.1.

Thanks to (5.13), the space $[A_{00}, A_{10}]_\theta$ is characterized with the help of the following theorem:

Theorem 5.2. *We have:*

$$[E_j(\Sigma), (H^{m_j+1/2,(m_j+1/2)/m}(\Sigma))']_\theta$$
$$= \mathscr{H}^{3m(1-\theta)-(m_j+1/2),[3m(1-\theta)-(m_j+1/2)]/m}(\Sigma)$$

if

$$\frac{3m(1-\theta) - (m_j + \frac{1}{2})}{m} < \frac{1}{2}. \tag{5.16}$$

(For the definition of the spaces $\mathscr{H}^{\alpha,\alpha/m}(\Sigma)$ see Chapter 4, Section 14.1).

Proof. 1) We first show that

$$[E_j(\Sigma), (H^{m_j+1/2,(m_j+1/2)/m}(\Sigma))']_{\theta_0} = L^2(\Sigma) \tag{5.17}$$
$$\text{if } 3m(1-\theta_0) - (m_j + \tfrac{1}{2}) = 0.$$

Indeed

$$H_{,0}^{3m-(m_j+1/2),[3m-(m_j+1/2)]/m}(\Sigma) \subset E_j(\Sigma) \subset$$
$$\subset H^{3m-(m_j+1/2),[3m-(m_j+1/2)]/m}(\Sigma),$$

so that $[E_j(\Sigma), (H^{m_j+1/2,(m_j+1/2)/m}(\Sigma))']_{\theta_0}$ contains and is contained in $L^2(\Sigma)$ (apply Theorems 14.2 and 14.3 of Chapter 4), from which (5.17) follows.

2) Next, we apply the *reiteration theorem* (Chapter 1, Section 6). If $\theta > \theta_0$, then:

$$[E_j(\Sigma), (H^{m_j+1/2,(m_j+1/2)/m}(\Sigma))']_\theta$$
$$= [[E_j(\Sigma), (H^{m_j+1/2,(m_j+1/2)/m}(\Sigma))']_{\theta_0}, (H^{m_j+1/2,(m_j+1/2)/m}(\Sigma))']_{\theta_1}$$

where

$$\theta = \theta_0 + \theta_1(1 - \theta_0);$$

according to (5.17), this space is equivalent to

$$[L^2(\Sigma), (H^{m_j+1/2,(m_j+1/2)/m}(\Sigma))']_{\theta_1},$$

which, by duality, reduces to (Chapter 1, Section 6.2)

$$[L^2(\Sigma), H^{m_j+1/2,(m_j+1/2)/m}(\Sigma)]_{\theta_1},$$

from which the result follows (by an application of Chapter 4, Section 2).

Now if $\theta < \theta_0$ (with or without (5.16)), then:

$$[E_j(\Sigma), (H^{m_j+1/2,(m_j+1/2)/m}(\Sigma))']_\theta$$
$$= [E_j(\Sigma), [E_j(\Sigma), (H^{m_j+1/2,(m_j+1/2)/m}(\Sigma))']_{\theta_0}]_{\theta_2},$$

where

$$\theta = \theta_0 \theta_2$$

and, by (5.17), this is equivalent to:

$$[E_j(\Sigma), L^2(\Sigma)]_{\theta_2}. \tag{5.18}$$

It remains, therefore, to characterize this space, in particular when (5.16) holds.

3) The space $E_j(\Sigma)$ appears as an *intersection*: define

$$F_j(\Sigma) = L^2(0, T; H^{3m-(m_j+1/2)}(\Gamma)), \tag{5.19}$$

$$G_j(\Sigma) = \Big\{ g \mid g \in H^{[3m-(m_j+1/2)]/m}(0, T; L^2(\Gamma)), \tag{5.20}$$

$$g(0) = 0,$$

and

$$g'(0) = 0, \quad \text{if} \quad \frac{3m - (m_j + \frac{1}{2})}{m} > \frac{3}{2},$$

or

$$\frac{1}{\sqrt{t}} g'(t) \in L^2(0, T; L^2(\Gamma)), \quad \text{if} \quad \frac{3m - (m_j + \frac{1}{2})}{m} = \frac{3}{2} \Big\}^{((1))}$$

Then

$$E_j(\Sigma) = F_j(\Sigma) \cap G_j(\Sigma). \tag{5.21}$$

((1)) See (11.52), Chapter 1.

But

$$F_j(\Sigma) \subset L^2(\Sigma), \quad G_j(\Sigma) \subset L^2(\Sigma),$$

and therefore

$F_j(\Sigma)$ = domain in $L^2(\Sigma)$ of a positive self-adjoint operator, say $\Phi_j = \Phi$

and

$G_j(\Sigma)$ = domain in $L^2(\Sigma)$ of a positive self-adjoint operator, say $\Psi_j = \Psi$.

Furthermore Φ and Ψ commute. (Indeed, after local maps and Fourier transformation, Φ amounts to multiplication by

$$(1 + |\xi|)^{3m-(m_j+1/2)}, \quad \xi \in \mathbf{R}^{n-1}$$

and Ψ to multiplication by

$$(1 + |\tau|)^{[3m-(m_j+1/2)]/m}, \tau \in \mathbf{R}).$$

Therfore (Chapter 1, Section 13):

$$[E_j(\Sigma), L^2(\Sigma)]_{\theta_2} = [F_j(\Sigma), L^2(\Sigma)]_{\theta_2} \cap [G_j(\Sigma), L^2(\Sigma)]_{\theta_2}. \tag{5.22}$$

But

$$[F_j(\Sigma), L^2(\Sigma)]_{\theta_2} = L^2(0, T; H^{(3m-(m_j+1/2))(1-\theta_2)}(\Gamma)) \tag{5.23}$$

and therefore the desired result is a consequence of the identity (see Chapter 4, (14.32)):

$$[G_j(\Sigma), L^2(\Sigma)]_{\theta_2} = H^{(3m-(m_j+1/2))(1-\theta_2)/m}(\Sigma) \tag{5.24}$$

if

$$\left(\frac{3m - (m_j + \frac{1}{2})}{m}\right)(1 - \theta_2) < \frac{1}{2}.$$

$\Bigg($Note that $\left(\frac{3m - (m_j + \frac{1}{2})}{m}\right)(1 - \theta_2)$

$$= \frac{3m - (m_j + \frac{1}{2})}{m}\left(1 - \frac{\theta}{\theta_0}\right) = \frac{3m(1-\theta) - (m_j + \frac{1}{2})}{m}\Bigg). \quad \square$$

5.3 Consequences

From (5.15) and Theorems 5.1 and 5.2, we deduce:

Theorem 5.3. *Assume that* $0 < \theta < 1$ *and that* (5.16) *holds for* $j = 0, \ldots, m-1$.

Let f, $\vec{g}$ *be given with*

$$f \in [H^{0,1}(Q), (H^{2m,2}(Q))']_\theta, \tag{5.25}$$

$$\vec{g} \in \prod_{j=0}^{m-1} \mathscr{H}^{3m(1-\theta)-(m_j+1/2),(3m(1-\theta)-(m_j+1/2))/m}(\Sigma). \tag{5.26}$$

Then, the solution u *of* (4.16) (with $u_0 = 0$, $u_1 = 0$)

$$(5.27)\quad \begin{cases} \textit{belongs to the space described in } (5.6), \textit{ or } (5.7) \\ \textit{if } \theta = \dfrac{1}{2}, \quad \textit{or } (5.8) \textit{ if } \theta = \dfrac{5}{6}. \end{cases}$$

Remark 5.1. The condition on θ is equivalent to

$$(5.28)\quad \theta > \frac{5}{6} - \frac{1}{6m} - \frac{\min m_j}{3m}. \quad \square$$

Remark 5.2. Formally, u is a solution of

$$(5.29)\quad \begin{cases} A\,u + u'' = f, \\ \vec{B}\,u = g \\ u(x,0) = 0, \quad u'(x,0) = 0. \quad \square \end{cases}$$

5.4 The Case of the Dirichlet Problem

For the Dirichlet problem, we may use Theorem 3.2 instead of the regularity Theorem 3.1.

We introduce the space $\mathscr{A}_0$ (which "replaces" the space A_0 defined in (5.2)):

$$(5.30)\quad \begin{cases} \mathscr{A}_0 = \Big\{ \vec{g}, u_0, u_1 \mid \vec{g} \in \prod_{j=0}^{m-1} H^{2m-j-1/2,(2m-j-1/2)/m}(\Sigma), \\ u_0 \in H^{3m/2}(\Sigma), \quad u_1 \in H^{m/2}(\Omega), \\ \vec{g}, u_0, u_1 \text{ satisfying the } \mathscr{R}.\mathscr{C}. \text{ defined by (3.30) to (3.33)} \Big\}. \end{cases}$$

If $\mathscr{G}$ denotes the mapping

$$f, \vec{g}, u_0, u_1 \xrightarrow{\mathscr{G}} u,$$

where u is a solution of

$$(5.31)\quad \begin{cases} A\,u + u'' = f, \\ \dfrac{\partial^j u}{\partial \nu^j} = g_j, \quad 0 \leqq j \leqq m-1, \\ u(x,0) = u_0(x), \dfrac{\partial u}{\partial t}(x,0) = u_1(x), \end{cases}$$

we now have:

$$(5.32)\quad \begin{cases} \mathscr{G} \text{ is a continuous mapping of } L^2(Q) \times \mathscr{A}_0 \to H^{m,1}(Q), \\ \mathscr{G} \text{ is a continuous mapping of } (H^{2m,2}(Q))' \times A_1 \to H^{0,-1}(Q). \end{cases}$$

Therefore, by interpolation of (5.32):

$$(5.33)\quad \begin{cases} \mathscr{G} \text{ is a continuous mapping of } [L^2(Q), ((H^{2m,2}(Q))']_\theta \times \\ \times [\mathscr{A}_0, A_1]_\theta \to [H^{m,1}(Q), H^{0,-1}(Q)]_\theta . \quad \square \end{cases}$$

As for Theorem 5.1, we can show:

Theorem 5.4. *We have the identities:*

$$(5.34)\quad \begin{cases} [H^{m,1}(Q), H^{0,-1}(Q)]_\theta = H^{-\theta}(0,T; H^{(1-\theta)m}(\Omega)) \cap \\ \qquad\qquad \cap H^{1-2\theta}(0,T; H^0(\Omega)), \\ \theta \neq \tfrac{1}{2}, \quad \theta \neq \tfrac{3}{4}; \end{cases}$$

$$(5.35)\quad [H^{m,1}(Q), H^{0,-1}(Q)]_{1/2} = H_{00}^{-1/2}(0,T; H^{m/2}(\Omega)) \cap H^0(Q);$$

$$(5.36)\quad [H^{m,1}(Q), H^{0,-1}(Q)]_{3/4} = H^{-3/4}(0,T; H^{m/4}(\Omega)) \cap \cap H_{00}^{-1/2}(0,T; H^0(\Omega)). \quad \square$$

By the same arguments as for Theorem 5.3 (introduce $\mathscr{A}_{00}$ = space of $\vec{g}$'s, with $u_0 = u_1 = 0$,

$$\vec{g} \in \prod_{j=0}^{m-1} H^{2m-j-1/2,(2m-j-1/2)/m}(\Sigma),$$

$$g_j(0) = 0 \quad \forall j, \quad g'(0) = 0 \quad \text{for } j > \frac{m-1}{2} \quad \text{and, for } j = \frac{m-1}{2},$$

$$\int_{\mathbf{R}_{x'}^{n-1}} \int_0^\infty |g'_{(m-1)/2}(x', \sigma^{2m})|^2 \, dx' \frac{d\sigma}{\sigma} < \infty \quad \text{if} \quad \Omega = \{x \mid x_n > 0\})$$

we obtain:

Theorem 5.5. *Let f be given with*

$$(5.37)\quad f \in [L^2(Q), (H^{2m,2}(Q))']_\theta$$

and let

$$(5.38)\quad \vec{g} \in \prod_{j=0}^{m-1} \mathscr{H}^{2m(1-\theta)-(j+1/2),(2m(1-\theta)-(j+1/2))/m}(\Sigma).$$

Assume that $u_0 = 0$ and $u_1 = 0$.

Assume that

$$(5.39)\quad \theta > \frac{3m-1}{4m}.$$

Then

(5.40) *if $\theta \neq \frac{3}{4}$, u belongs to the space described in* (5.34);

(5.41) *if $\theta = \frac{3}{4}$, u belongs to the space described in* (5.36). $\square$

Remark 5.3. It is interesting to *compare* Theorem 5.5 with Theorem 5.3 applied to the Dirichlet problem, that is with $m_j = j$, $\forall j$. Let us just concern ourselves with the regularity of the g_j's. Take $\theta = \theta_1$ in Theorem 5.3 and $\theta = \theta_2$ in Theorem 5.5. Then, we have the same regularity for the g_j's if

$$3(1 - \theta_1) = 2(1 - \theta_2).$$

Then $2 - 3\theta_1 = 1 - 2\theta_2$. Therefore, Theorems 5.3 and 5.5 yield respectively:

$$u \in H^{-\theta_1}(0, T; H^{2m(1-\theta_1)}(\Omega)) \cap H^{2-3\theta_1}(0, T; H^0(\Omega))$$

and

$$u \in H^{-2\theta_2}(0, T; H^{(1-\theta_2)m}(\Omega)) \cap H^{2-3\theta_1}(0, T; H^0(\Omega)). \quad \square$$

Remark 5.4. For the Dirichlet problem, we can also transpose the isomorphism $\varphi \to v$, wehre v is the solution of

$$A^* v + v'' = \varphi,$$

$$v(x, T) = 0, \quad v'(x, T) = 0,$$

$$\frac{\partial^j v}{\partial \nu^j} = 0 \quad \text{on} \quad \Sigma, \quad 0 \leqq j \leqq m - 1,$$

for $\varphi \in L^2(Q)$. $\square$

6. Applications and Examples

6.1 General Results

We consider problem (5.29), assuming that the data g_j satisfy

(6.1) $$g_j \in L^2(\Sigma), \quad j = 0, 1, \ldots, m - 1.$$

In order to simplify, we choose

(6.2) $$u_0 = 0, \quad u_1 = 0.$$

Our aim is to investigate (assuming f to be chosen in an "appropriate" manner) the corresponding order of regularity for the solution u. $\square$

Case in which the data are not of Dirichlet type.

We apply Theorem 5.3. This is permissible for θ such that

$$\mathscr{H}^{3m(1-\theta)-(m_j+1/2),(3m(1-\theta)-(m_j+1/2))/m}(\Sigma) \supset L^2(\Sigma), \quad \forall j,$$

that is if

(6.3) $$3m(1 - \theta) - (m_j + \tfrac{1}{2}) \leqq 0, \quad \forall j.$$

The optimal choice of θ (i.e. which yields the maximum information about the regularity of the corresponding solution u) is the least θ such that (6.3) is satisfied, that is

$$\theta = 1 - \frac{1}{6m} - \frac{\min m_j}{3m} \tag{6.4}$$

Let us examine the particular cases:

(i) the case $\theta = \frac{1}{2}$ yields: $2 \min m_j = 3m - 1$; but $\min m_j \leqq m$ and therefore this is possible only if $m = 1$, $m_0 = 1$;

(ii) the case $\theta = \dfrac{5}{6}$ yields: $\min m_j = \dfrac{m-1}{2}$, which is possible if m is odd.

In conclusion:

Theorem 6.1. *Let the g_j's be given in $L^2(\Sigma)$ and f be given by* (5.25), *with θ given by* (6.4). *Assume that $u_0 = 0$, $u_1 = 0$. Then, unless*

$$m = 1, \quad m_0 = 1, \tag{6.5}$$

or

$$m \textit{ is odd}, \quad \min m_j = \frac{m-1}{2}, \tag{6.6}$$

we have

$$u \in H^{-\theta}(0, T; H^{2m(1-\theta)}(\Omega)) \cap H^{2-3\theta}(0, T; H^0(\Omega)). \tag{6.7}$$

For the case (6.5), *we have*

$$u \in H_{00}^{-1/2}(0, T; H^1(\Omega)) \cap H^{1/2}(0, T; H^0(\Omega)). \tag{6.8}$$

For the case (6.6), *we have*

$$u \in H^{-5/6}(0, T; H^{m/3}(\Omega)) \cap H_{00}^{-1/2}(0, T; H^0(\Omega)). \quad \square \tag{6.9}$$

The case in which the data are of Dirichlet type.

Here, we apply Theorem 5.5. This is legitimate if

$$\mathscr{H}^{2m(1-\theta)-(j+1/2)(2m(1-\theta)-(j+1/2))/m}(\Sigma) \supset L^2(\Sigma), \quad \forall j,$$

that is, if

$$2m(1-\theta) - \tfrac{1}{2} \leqq 0$$

and therefore the optimal θ is

$$\theta = 1 - \frac{1}{4m}. \tag{6.10}$$

The only singular case, $\theta = \frac{3}{4}$, corresponds to $m = 1$. Therefore:

Theorem 6.2. *Let the g_j's be given in $L^2(\Sigma)$ and f be given by* (5.37), *with θ given by* (6.10). *Then, unless*

$$m = 1, \tag{6.11}$$

we have

$$(6.12)\qquad u \in H^{-\theta}(0, T; H^{(1-\theta)m}(\Omega)) \cap H^{1-2\theta}(0, T; H^0(\Omega)).$$

For the case (6.11), *we have*

$$(6.13)\qquad u \in H^{-3/4}(0, T; H^{1/4}(\Omega)) \cap H_{00}^{-1/2}(0, T; H^0(\Omega)). \quad \square$$

6.2 Examples

Example 6.1

Consider the problem

$$(6.14)\qquad \begin{cases} -\Delta u + u'' = 0 \\ \dfrac{\partial u}{\partial \nu} = g \quad \text{on} \quad \Sigma, \\ u(x, 0) = 0, \quad u'(x, 0) = 0. \end{cases}$$

Then $m = 1$, $m_0 = 1$; therefore we have the case (6.5) of Theorem 6.1; we obtain:

$$(6.15)\qquad \begin{cases} \text{if } g \in L^2(\Sigma), \text{ the solution } u \text{ of (6.14) satisfies} \\ u \in H_{00}^{-1/2}(0, T; H^1(\Omega)) \cap H^{1/2}(0, T; H^0(\Omega)). \end{cases}$$

Remark 6.1. We shall see how result (6.15) compares to the result given by Theorem 9.5 of Chapter 3.

First of all, let us see how the preceding problem may be stated in *variational form*. Introduce

$$V = H^1(\Omega), \qquad H = L^2(\Omega)$$

and $V' = V^{-1}$ = antidual of V (it is not a space of distributions on Ω). For $v \in H^{1/2+\varepsilon}(\Omega)$, arbitrary $\varepsilon > 0$, we know (Chapter 1, Section 8) that $\gamma_0 v$ = trace of v on $\Gamma \in H^{\varepsilon}(\Gamma)$. Consequently, for g given in $L^2(\Sigma) = L^2(0, T; L^2(\Gamma))$, we may define f by:

$$(6.16)\qquad (f(t), v) = \int_\Gamma g(x, t)\, \overline{\gamma_0 v(x)}\, d\Gamma$$

and thus

$$(6.17)\qquad f \in L^2(0, T; (H^{1/2+\varepsilon}(\Omega))'), \quad \forall \varepsilon > 0.$$

But (6.17) is equivalent (by Chapter 1, Section 9) to

$$(6.18)\qquad \begin{cases} f \in L^2(0, T; V^{-1/2-\varepsilon}) \\ V^{-\theta} = [H, V']_\theta. \end{cases}$$

Thus, the problem is *equivalent* to the search for the solution u of

$$a(u(t), v) + (u''(t), v) = (f(t), v), \quad \forall v \in V, \tag{6.19}$$

where

$$a(u, v) = \sum_{i=1}^{n} \int_{\Omega} \frac{\partial u}{\partial x_i} \overline{\frac{\partial v}{\partial x_i}} \, dx$$

and where (,) denotes the scalar product in H (or in the antiduality between V and V').

Therefore, Theorem 9.5 of Chapter 3 applies, with

$$\theta = \tfrac{1}{2} + \varepsilon, \quad \text{arbitrary } \varepsilon > 0.$$

We obtain:

$$u \in C^0([0, T]; V^{1/2-\varepsilon}), \quad \forall \varepsilon > 0, \tag{6.20}$$

$$u' \in C^0([0, T]; V^{-1/2-\varepsilon}), \quad \forall \varepsilon > 0. \tag{6.21}$$

Therefore, *in particular:*

$$u \in L^2(0, T; V^{1/2-\varepsilon}), \quad u' \in L^2(0, T; V^{-1/2-\varepsilon})$$

and therefore, according to Chapter 1, Section 4:

$D_t^{1/2} u$ ($\tfrac{1}{2}$-order derivative of u with resprect to t) $\in L^2(0, T; V^{-\varepsilon})$.

In summary:

$$u \in L^2(0, T; H^{1/2-\varepsilon}(\Omega)) \cap H^{1/2}(0, T; V^{-\varepsilon}), \quad \forall \varepsilon > 0. \tag{6.22}$$

It can be shown that (6.15) implies

$$u \in L^2(0, T; [H^1(\Omega), H^0(\Omega)]_{1/2}) = L^2(0, T; H^{1/2}(\Omega)),$$

so that *we may take* $\varepsilon = 0$ *in* (6.22) — a result which does not follow from Theorem 9.5, Chapter 3.

On the other hand, the results of this chapter, which are all of a "Hilbert" nature, *do not yield* (6.20) and (6.21). □

Remark 6.2. In all this chapter, *no* regularity result is optimal, since from the start we do not have optimal regularity results. This may be verified with a trivial example; take (6.14) in one dimension, with $\Omega =]0, \infty[$ (the fact that Ω is unbounded is of course without significance for the study of a *local* regularity result). Then

$$u(x, t) = \left\{ \int_0^{t-x} g(\sigma)\, d\sigma, \quad \text{if} \quad t \geqq x;\ 0, \text{ if } t \leqq x \right\};$$

the solution is locally in H^1 in the neighborhood of Σ; which is of course better than the result given by (6.15). □

Example 6.2

Now, consider

$$(6.23)\qquad \begin{cases} -\Delta u + u'' = 0 \\ u = g \quad \text{on} \quad \Sigma, \\ u(x,0) = 0, \qquad u'(x,0) = 0. \end{cases}$$

We can apply:

Theorem 6.1, with $m = 1$, case (6.6); Theorem 6.2, with $m = 1$.

Then:

if $g \in L^2(\Sigma)$, the solution u of (6.23) satisfies

$$(6.24)\qquad u \in H^{-5/6}(0,T; H^{1/3}(\Omega)) \cap H^{-3/4}(0,T; H^{1/4}(\Omega)) \cap \cap H_{00}^{-1/2}(0,T; H^0(\Omega)). \quad \square$$

Remark 6.3. Same remark as Remark 6.2. Taking $\Omega =]0, \infty[$, this time we obtain

$$u(x,t) = \{g(t-x) \text{ if } t > x;\ 0 \text{ if } t < x\},$$

a function which is locally in L^2 in the neighborhood of Σ (which is better than (6.24)). □

Example 6.3

Consider the problem

$$(6.25)\qquad \begin{cases} \Delta^2 u + u'' = 0, \\ \Delta u|_\Sigma = g_0, \quad \left.\dfrac{\partial \Delta u}{\partial \nu}\right|_\Sigma = g_1, \\ u(x,0) = 0, \quad u'(x,0) = 0. \end{cases}$$

Then, $m = 2$, $m_0 = 2$, $m_1 = 3$, $\min m_j = 2$.

Formula (6.4) yields $\theta = \frac{3}{4}$ and we obtain:

$$(6.26)\qquad u \in H^{-3/4}(0,T; H^1(\Omega)) \cap H^{5/4}(0,T; H^0(\Omega)). \quad \square$$

Example 6.4

Consider the problem

$$(6.27)\qquad \begin{cases} \Delta^2 u + u'' = 0, \\ u|_\Sigma = g_0, \quad \left.\dfrac{\partial u}{\partial \nu}\right|_\Sigma = g_1, \\ u(x,0) = 0, \quad u'(x,0) = 0. \end{cases}$$

Then $m = 2$, $\min m_j = 0$ and Theorem 6.1 yields

$$u \in H^{-11/12}(0, T; H^{1/3}(\Omega)) \cap H^{-3/4}(0, T; H^0(\Omega)) \tag{6.28}$$

and Theorem 6.2 yields:

$$u \in H^{-7/8}(0, T; H^{1/4}(\Omega)) \cap H^{-3/4}(0, T; H^0(\Omega)). \quad \square \tag{6.29}$$

7. Regularity Theorem (II)

7.1 Statement

We again consider the problem:

$$A\,u + u'' = f \tag{7.1}$$

$$\vec{B}\,u = 0, \tag{7.2}$$

$$u(x, 0) = 0, \quad u'(x, 0) = 0. \tag{7.3}$$

We shall prove the following *regularity theorem:*

Theorem 7.1. *Let r be an integer $\geqq 0$. Let f be given with*

$$\begin{cases} f \in L^2(0, T; H^{(2r-1)m}(\Omega)), \\ f^{(2r)} = \dfrac{d^{2r}}{dt^{2r}} f \in L^2(0, T; H^0(\Omega)) = L^2(Q), \end{cases} \tag{7.4}$$

and

$$\begin{cases} f(x, 0) = f'(x, 0) = \cdots = f^{(2r-1)}(x, 0) = 0 \\ \text{(no condition if } r = 0.) \end{cases} \tag{7.5}$$

Then the solution u of (7.1), (7.2), (7.3) *satisfies*

$$u \in H^{(2r+1)m, 2r+1}(Q) \tag{7.6}$$

(that is:

$$u \in L^2(0, T; H^{(2r+1)m}(\Omega)), \quad u^{(2r+1)} \in L^2(0, T; H^0(\Omega))).$$

Remark 7.1. If $r = 0$, (7.4) reduces to

$$f \in L^2(Q).$$

Then (7.6) holds (Chapter 3). $\square$

7.2 Proof of Theorem 7.1

1) By formal differentiation in t, we have

$$(7.7)\qquad A\,u^{(2r)} + u^{(2r+2)} + \sum_{k=0}^{2r-1} \binom{2r}{k} A^{(2r-k)}\,u^{(k)} = f^{(2r)}$$

$$\left(\text{where } A^{(j)}\varphi = \sum_{|p|,|q|\leq m} (-1)^{|p|} D_x^p(D_t^j\,a_{pq}(x,t))\,D_x^p\varphi\right)$$

and

$$(7.8)\qquad u^{(j)}(0) = 0, \quad 0 \leqq j \leqq 2r+1.$$

Multiplying (7.7) with $u^{(2r+1)}(t)$ and applying the appropriate integrations by parts (as in Section 1), we obtain

$$\|u^{(2r+1)}(t)\|^2_{H^0(\Omega)} + \|u^{(2r)}(t)\|^2_{H^m(\Omega)} \leqq$$

$$\leqq c\int_0^t [\|u^{(2r)}(\sigma)\|^2_{H^m(\Omega)} + \|u^{(2r+1)}(\sigma)\|^2_{H^0(\Omega)} + \|f^{(2r)}(\sigma)\|^2_{H^0(\Omega)}]\,d\sigma$$

from which, by an application of Gronwall's lemma, we obtain an a priori *inequality* of the form

$$(7.9)\qquad \|u^{(2r+1)}(t)\|^2_{H^0(\Omega)} + \|u^{(2r)}(t)\|^2_{H^m(\Omega)} \leqq c\int_0^T \|f^{(2r)}(\sigma)\|^2_{H^0(\Omega)}\,d\sigma.$$

In particular, it follows (using, for instance, the method of Faedo-Galerkin) that

$$(7.10)\qquad u^{(2r)} \in L^2(0,T;H^m(\Omega)),$$

$$(7.11)\qquad u^{(2r+1)} \in L^2(0,T;H^0(\Omega)).$$

2) We shall prove the theorem by induction on r. It holds for $r=0$ (see Remark 7.1). If we admit that it holds up to $(r-1)$, then we obtain

$$(7.12)\qquad u \in L^2(0,T;H^{(2r-1)m}(\Omega)) \text{ (and also } u^{(2r-1)} \in L^2(0,T;H^0(\Omega))),$$

which, together with (7.11) and the intermediate derivative theorem (Chapter 1, Section 2), yields

$$(7.13)\qquad u^{(j)} \in L^2(0,T;H^{(2r-1)m(1-j/(2r+1))}(\Omega)), \quad 0 \leqq j \leqq 2r+1.$$

We shall, as a first step, show that

$$(7.14)\qquad u \in L^2(0,T;H^{(2r+1)m-\varepsilon}(\Omega)) \text{ for all } \varepsilon > 0.$$

Set

$$(7.15)\qquad \xi_0 = (2r-1)\,m, \quad \xi_n = \xi_{n-1}\left(1 - \frac{2}{2r+1}\right) + 2m, \quad n \geqq 1;$$

according to (7.13):

$$u' \in L^2(0, T; H^{\xi_0(1-2/(2r+1))}(\Omega));$$

and since f satisfies (7.4), we have:

$$f \in L^2(0, T; H^{\xi_0(1-2/(2r+1))}(\Omega)),$$

since

$$\xi_0 \left(1 - \frac{2}{2r+1}\right) < (2r - 1)\, m.$$

Therefore

$$A\, u = f - u'' \in L^2(0, T; H^{\xi_0(1-2/(2r+1))}(\Omega)),$$

$$\vec{B}\, u = 0;$$

from which, as in Lemma 2.1, we deduce:

$$u \in L^2(0, T; H^{\xi_1}(\Omega)).$$

Admitting, by induction on n, that

$$u \in L^2(0, T; H^{\xi_{n-1}}(\Omega)),$$

we obtain (with (7.11) and the intermediate derivative theorem):

$$u'' \in L^2(0, T; H^{\xi_{n-1}(1-2/(2r+1))}(\Omega)).$$

But

$$\xi_{n-1} \left(1 - \frac{2}{2r+1}\right) < (2r - 1)\, m,$$

therefore

$$A\, u = f - u'' \in L^2(0, T; H^{\xi_{n-1}(1-2/(2r+1))}(\Omega)),$$

$$\vec{B}\, u = 0;$$

therefore

(7.16) $$u \in L^2(0, T; H^{\xi_n}(\Omega)).$$

We therefore have (7.16) for all n. But $\xi_n \to \xi = (2r + 1)\, m$, whence (7.14).

3) We shall now show, *by induction on* k, integer k increasing, that

(7.17) $$u^{(2(r-k))} \in L^2(0, T; H^{(2k+1)m}(\Omega)), \quad k \leqq r.$$

The result is known for $k = 0$. We assume it to hold up to $(k - 1)$, that is

(7.18) $$u^{(2(r-k)+2)} \in L^2(0, T; H^{(2k-1)m}(\Omega)).$$

Differentiate (7.1) $2(r-k)$ times with respect to t:

$$(7.19)\quad A\,u^{(2(r-k))} + u^{(2(r-k)-r)} + \sum_{j=0}^{2(r-k)-1} A^{(2(r-k)-j)} u^{(j)} = f^{(2(r-k))}.$$

According to (7.14) and (7.11),

$$u^{(j)} \in L^2(0, T; H^{(2r+1)m(1-j/(2r+1))-\xi_j}(\Omega)),$$

arbitrary $\xi_j > 0$, and

$$(2r+1)\,m\left(1 - \frac{j}{2r+1}\right) > (2k+1)\,m \quad \text{if} \quad j \leqq 2(r-k) - 1,$$

therefore, in (7.19):

$$(7.20)\qquad A^{((2r-k)-j)}\, u^{(j)} \in L^2(0, T; H^{(2k-1)m}(\Omega)).$$

Furthermore

$$f^{(2(r-k))} \in L^2(0, T; H^{(2r-1)m(1-2(r-k)/2r)}(\Omega))$$

and

$$(2r-1)\,m\left(1 - \frac{2(r-k)}{2r}\right) \geqq (2k-1)\,m,$$

which, together with (7.18), (7.20), and (7.19) yields

$$A\,u^{(2(r-k))} \in L^2(0, T; H^{(2k-1)m}(\Omega)),$$

$$\vec{B}\,u = 0$$

and therefore we have (7.17), which proves the theorem. □

Remark 7.2. If A is independent of t and if we make somewhat stronger hypotheses on f (see below), we can give another proof of the preceding theorem – proof which shows that (7.6) is essentially the best possible result (at least in the Hilbert space setting).

We assume the hypotheses of Section 1.3 to be satisfied and use the same notation. We introduce

$$(7.21)\qquad \hat{u}(\lambda, t) = (\mathscr{U}\,u(t))\,(\lambda), \quad \hat{f}(\lambda, t) = (\mathscr{U}\,f(t))\,(\lambda)$$

and assume

$$(7.22)\qquad \begin{cases} \lambda^{r-1/2}\,\hat{f} \in L^2(0, T; \mathfrak{h}), \\ \dfrac{d^{2r}}{dt^{2r}}\hat{f} \in L^2(0, T; \mathfrak{h}), \end{cases}$$

and that (7.5) holds.

The first condition in (7.22) is *more restrictive* than the first condition in (7.4): it means that $f \in L^2(0, T; D(A^{r-1/2}))$, which implies that

$$f \in L^2(0, T; H^{(2r-1)m}(\Omega))$$

and *furthermore* that f satisfies certain boundary conditions on Σ.

We shall recover (7.6) by applying the explicit formula:

$$\hat{u} = E \underset{(t)}{*} f, \quad E(\lambda, t) = \begin{cases} \dfrac{1}{\sqrt{\lambda}} \sin\left(t\sqrt{\lambda}\right), & t > 0 \\ 0, & t < 0. \end{cases} \tag{7.23}$$

From the formula

$$\frac{d^2 E}{dt^2} + \lambda E = \delta(= \delta(t)),$$

we deduce that

$$E = \frac{1}{\lambda}\left[\delta - \frac{1}{\lambda}\delta'' + \frac{1}{\lambda^2}\delta^{(4)} + \cdots + (-1)^{r-1}\frac{1}{\lambda^{r-1}}\delta^{(2r-2)} + \right. \\ \left. + (-1)^r \frac{1}{\lambda^{r-1}} \frac{d^{2r}}{dt^{2r}} E\right]. \tag{7.24}$$

Since (7.5) holds, the derivatives of f on $\mathbf{R}_t$ do not introduce any mass at the t-origin and therefore, introducing (7.24) into (7.23), we obtain

$$\hat{u} = \frac{1}{\lambda} f - \frac{1}{\lambda^2} f'' + \cdots + (-1)^{r-1}\frac{1}{\lambda^r} f^{(2r-2)} + \\ + (-1)^r \frac{1}{\lambda^r}(E * f^{(2r)}). \tag{7.25}$$

We must show that

$$\lambda^{(r+1/2)}\, \hat{u} \in L^2(0, T; \mathfrak{h}) \tag{7.26}$$

(which implies (7.6) and in fact, in this case, is equivalent to (7.6), since the boundary conditions for u are automatically satisfied).

But $\lambda^{(r+1/2)}\left(\dfrac{1}{\lambda} f\right)$ belongs to $L^2(0, T; \mathfrak{h})$ according to the first hypothesis in (7.22) and

$$\lambda^{r+1/2}\frac{1}{\lambda^r}(E * f^{(2r)}) = \lambda^{1/2}(E * f^{(2r)}) = \left(\sin t\sqrt{\lambda}\right) \underset{(t)}{*} f^{(2r)}$$

belongs to $L^2(0, T; \mathfrak{h})$ according to the second hypothesis in (7.22) — and this cannot be improved upon.

It remains to show that

(7.27) $$\lambda^{r+1/2}\frac{1}{\lambda^{j+1}}\hat{f}^{(2)} \in L^2(0, T; \mathfrak{h})$$

$$\text{for} \quad 1 \leqq j \leqq r-1.$$

Now, from (7.22) and the result on intermediate derivatives (Chapter 1, Section 3), we deduce that

$$\lambda^{(r-1/2)(1-2j/2r)}\hat{f}^{(2j)} \in L^2(0, T; \mathfrak{h}),$$

i.e.

$$\lambda^{r-1/2-j+j/2r}\hat{f}^{(2j)} \in L^2(0, T; \mathfrak{h}),$$

which, *in particular*, implies (7.27) [thus, we could improve the result on the "intermediate" terms in (7.25), but not on the extreme terms]. Hence, our assertion follows.

8. Non-Integer Order Regularity Theorem

8.1 Orientation

As we have seen in Chapter 4, Sections 8, 9 and 10, we can either transpose the situation of Section 6 *and then* interpolate, or interpolate *first* – and this second possibility is probably more favorable here.

We therefore briefly indicate the "interpolation in r" of Theorem 7.1.

8.2 Interpolation in r

Conditions (7.4) may be summarized by:

(8.1) $$f \in V^{(2r-1)m,2r}(Q),$$

where

(8.2) $$\begin{cases} V^{(2r-1)m,2r}(Q) = \{f \mid f \in H^{(2r-1)m,2r}(Q); \\ f(x,0) = \cdots = f^{(2r-1)}(x,0) = 0\}. \end{cases}$$

(the space $H^{2(r-1)m,2r}_{,0}(Q)$ (see Chapter 4, Section 2) consists of the elements $f \in V^{(2r-1)m,2r}(Q)$ such that, *furthermore*, $f(x,T) = \cdots = f^{(2r-1)}(x,T) = 0$).

According to Theorem 7.1,

(8.3) $$f \to u = \text{"solution of (7.1), (7.2), (7.3)"}$$

is a continuous linear mapping of

$$V^{(2r_0-1)m,2r_0}(Q) \to H^{(2r_0+1)m,2r_0+1}(Q), \quad \text{integer } r_0 > 1,$$

$$V^{m,2}(Q) \to H^{3m,3}(Q).$$

Define:

$$(8.4)\qquad \begin{cases} V^{(2r-1)m,2r}(Q) = [V^{(2r_0-1)m,2r_0}(Q),\, V^{m,2}(Q)]_\theta, \\ r \geqq 1 \ \text{ and arbitrary } (<r_0, \text{ but } r_0 \text{ is arbitrary!}), \end{cases}$$

where

$$(8.5)\qquad r = (1-\theta)\, r_0 + \theta .$$

(Proposition 8.1 will show that, up to an equivalence in norms, definition (8.4) depends only on r and not on r_0 and θ.)

Since (see Chapter 4, Section 2), with (8.5):

$$[H^{(2r_0+1)m,2r_0+1}(Q),\, H^{3m,3}(Q)]_\theta = H^{(2r+1)m,2r+1}(Q)$$

it follows that we have:

Theorem 8.1. *Let r be arbitrary, $r \geqq 1$. Then, if $f \in V^{(2r-1)m,2r}(Q)$ (defined by* (8.4) *and* (8.2)), *the solution u of* (7.1), (7.2), (7.3) *belongs to* $H^{(2r+1)m,2r+1}(Q)$. □

Remark 8.1. The case $0 \leqq r \leqq 1$ can be treated in the same way, using the fact that $f \to u$ is a continuous mapping of $H^0(Q) \to H^{m,1}(Q)$. □

8.3 Interpretation of the Space $V^{(2r-1)m,2r}(Q)$, $r \geqq 1$

Proposition 8.1. *For $2r \neq$ integer $+ \frac{1}{2}$, we have:*

$$V^{(2r-1)m,2r}(Q) = \{f \mid f \in H^{(2r-1)m,2r}(Q), \quad f^{(j)}(x,0) = 0 \ \text{ for } 0 \leqq j < 2r - \tfrac{1}{2}\}$$

(therefore definition (8.2) for *integer* r extends to the non-integer case).

Remark 8.2. If $2r =$ integer $+ \frac{1}{2}$, then we have an *integral relation* (see Chapter 4, Section 2.5). □

Proof. We have:

$$(8.6)\qquad V^{(2r_0-1)m,2r_0}(Q) = L^2(0,T; H^{(2r_0-1)m}(\Omega)) \cap G^{r_0}$$

where

$$(8.7)\qquad G^{r_0} = \{f \mid f \in H^{2r_0}(0,T; H^0(\Omega)),\, f^{(j)}(0) = 0,\, 0 \leqq j \leqq 2r_0 - 1\}$$

and similarly

$$(8.8)\qquad V^{m,2}(Q) = L^2(0,T; H^m(\Omega)) \cap G^1 .$$

But $L^2(0,T; H^{(2r_0-1)m}(\Omega))$ is the domain in $L^2(Q)$ of a positive self-adjoint operator Λ_0 and $L^2(0,T; H^m(\Omega)) = D(\Lambda_0^{\alpha_0})$ with:

$$(2r_0 - 1)\, \alpha_0 = 1 .$$

On the other hand, G^{r_0} is the domain in $L^2(Q)$ of a positive self-adjoint operator Λ_1 and $G^1 = D(\Lambda_1^{\alpha_1})$ with $r_0\, \alpha_1 = 1$ (this can be shown with the same type of technique as for the interpolation between $H_0^m(\Omega)$ and $H_0^1(\Omega)$; see Chapter 1). Therefore

$$V^{(2r-1)m,2r}(Q) = [D(\Lambda_0) \cap D(\Lambda_1), D(\Lambda_0^{\alpha_0}) \cap D(\Lambda_1^{\alpha_1})]_\theta$$

and since the operators Λ_0 and Λ_1 *commute*, this last space coincides with

$$[D(\Lambda_0), D(\Lambda_0^{\alpha_0})]_\theta \cap [D(\Lambda_1), D(\Lambda_1^{\alpha_1})]_\theta,$$

therefore:

$$V^{(2r-1)m,2r}(Q) = [L^2(0, T; H^{(2r_0-1)m}(\Omega)), L^2(0, T; H^m(\Omega))]_\theta \cap [G^{r_0}, G^1]_\theta. \tag{8.9}$$

The first space on the right side of the equality sign is $L^2(0, T; H^{(2r-1)m}(\Omega))$ and, as in Chapter 1, Section 11.5, the second space coincides with

$$G^r = \{f \mid f \in H^{2r}(0, T; H^0(\Omega)), f^{(j)}(0) = 0, 0 \leqq j < 2r - \tfrac{1}{2}\}$$

from which the result follows. □

9. Adjoint Isomorphism of Order r and Transposition

9.1 Adjoint Isomorphism of Order r

We shall use only Theorem 7.1, that is the case integer r – and furthermore $r \geqq 1$. The adaption to "non-integer r" corresponding to Section 8, is not given here. □

As usual (see Section 4.1), we consider the *adjoint problem:*

$$\begin{cases} A^* v + v'' = \varphi, \\ \tilde{C} v = 0, \\ v(T) = v'(T) = 0, \end{cases} \tag{9.1}$$

and we assume that

$$\varphi \in H_{0,0}^{(2r-1)m,2r}(Q) = \text{closure of } \mathcal{D}(Q) \text{ in } H^{(2r-1)m,2r}(Q). \tag{9.2}$$

Define:

$$X^r(Q) = \text{space described by } v \text{ as } \varphi \text{ describes } H_{0,0}^{2(r-1)m,2r}(Q) \tag{9.3}$$

and then ($X^r(Q)$ being provided with the "translated" topology) we have (done what was necessary to have):

$$v \to A^* v + v'' \text{ is an isomorphism of } X^r(Q) \to H_{0,0}^{(2r-1)m,2r}(Q). \tag{9.4}$$

This is what we shall call the *adjoint isomorphism of order r.*

9.2 Transposition

Let us transform (9.4). Note that (see Chapter 4, Section 8.3)

$$(9.5)\qquad (H^{(2r-1)m,2r}_{0,0}(Q))' = H^{-(2r-1)m,-2r}(Q);$$

and recall that every element g of this space may be written (non-uniquely):

$$(9.6)\qquad \begin{cases} g = \sum_{|p|\leq(2r-1)m} D_x^p g_p + \sum_{0\leq k\leq 2r} D_t^k h_k, \\ g_p, h_k \in L^2(Q). \quad \square \end{cases}$$

Then:

Theorem 9.1. *Let $v \to L(v)$ be a continuous antilinear form on $X^r(Q)$; there exists a unique u in $H^{-(2r-1)m,-2r}(Q)$ such that*

$$(9.7)\qquad \langle u, \overline{A^* v + v''}\rangle = L(v), \quad \forall v \in X^r(Q).$$

9.3 Formal Choice of L

As in (4.11), we choose

$$(9.8)\qquad L(v) = \langle f, \bar{v}\rangle + \langle \vec{g}, \overline{\vec{T} v}\rangle + \langle u_1, \overline{v(0)}\rangle - \langle u_0, \overline{v'(0)}\rangle,$$

but (which we have not done in Section 4), we choose f, $\vec{g}$, u_i to be (suitable) *distributions* on Q, Σ and Ω.

Thus, we shall follow a procedure analogous to Chapter 4, Sections 8 to 12. □

10. Choice of f, $\vec{g}$, u_0, u_1

10.1 Choice of f

According to Theorem 7.1, we have:

$$(10.1)\qquad X^r(Q) \subset H^{(2r+1)m,2r+1}(Q).$$

Introduce (as in Chapter 4, Section 9) the space:

$$(10.2)\qquad \Xi^{(2r+1)m,2r+1}(Q) = \{v \mid d^j(t)\, v^{(j)} \in L^2(0,T; \Xi^{(2r+1-j)m}(\Omega)),\ 0 \leq j \leq 2r+1\},$$

where

$d(t)$ is defined by (9.3) of Chapter 4,

$\Xi^\mu(\Omega)$ is defined in Chapter 2, Section 6.3.

As for Proposition 9.1 of Chapter 4, we verify that

$$(10.3)\qquad \mathscr{D}(Q) \text{ is dense in } \Xi^{(2r+1)m,2r+1}(Q).$$

Therefore

(10.4) $$\Xi^{-(2r+1)m,-(2r+1)}(Q) = (\Xi^{(2r+1)m,(2r+1)}(Q))' \subset \mathscr{D}'(Q).$$

For this space, we have a structure result analogous to Proposition 9.3 of Chapter 4 and, thanks to (10.1), we have:

(10.5) $$\text{if } f \in \Xi^{-(2r+1)m,-(2r+1)}(Q), \quad \text{then } v \to \langle f, v \rangle$$

is *continuous* antilinear on $X^r(Q)$.

10.2 The Space $D_{A+D_t^2}^{-(2r-1)}(Q)$

Taking, in (9.7), $v = \varphi \in \mathscr{D}(Q)$, we have $L(\varphi) = (f, \varphi)_Q$ and therefore, in the sense of $\mathscr{D}'(Q)$:

(10.6) $$A u + u'' = f.$$

This leads to the following *definition:*

(10.7) $$\begin{cases} D_{A+D_t^2}^{-(2r-1)}(Q) = \{ u \mid u \in H^{-(2r-1)m,-2r}(Q), \\ A u + u'' \in \Xi^{-(2r+1)m-(2r+1)}(Q)\}, \quad \text{integer } r \geqq 1; \end{cases}$$

provided with the norm of the graph, this is a Hilbert space.

Then the solution u of (9.7) belongs to $D_{A+D_t^2}^{-(2r-1)}(Q)$.

10.3 Choice of g_j

From (10.1) and the trace theorems of Chapter 1 (see also Chapter 4, Section 2), we deduce that:

(10.8) $$\begin{cases} v \to \vec{T} v \quad \text{is a continuous linear mapping of} \\ X^r(Q) \to \prod_{j=0}^{m-1} H^{(2r+1)m-(2m-m_j-1/2),((2r+1)m-(2m-m_j-1/2))/m}(\Sigma). \end{cases}$$ □

We introduce (compare with Chapter 4, Section 11.1) the spaces

$$H^{\alpha m} \Xi^{\alpha}(\Sigma), \quad \text{real } \alpha \geqq 0;$$

first for *integer* $\alpha \geqq 0$:

(10.9) $$H^{\alpha m} \Xi^{\alpha}(\Sigma) = \{v \mid d^j(t)\, v^{(j)} \in L^2(0, T; H^{(\alpha-j/\alpha)m}(\Gamma)), 0 \leqq j \leqq \alpha\}$$

and then for non-integer $\alpha \in \mathbf{R}_+$, we define by interpolation:

(10.10) $$\begin{cases} H^{\alpha m} \Xi^{\alpha}(\Sigma) = [H^{\alpha_0 m} \Xi^{\alpha_0}(\Sigma), H^{0,0}(\Sigma)]_{\theta} \\ \text{integer } \alpha_0, \quad (1-\theta)\,\alpha_0 = \alpha. \end{cases}$$

The space defined in this way depends only on α. Then:

$$(10.11)\quad \begin{cases} H^{(2r-1)m+m_j+1/2,((2r-1)m+m_j+1/2)/m}(\Sigma) \subset \\ \subset H^{(2r-1)m+m_j+1/2}\,\Xi^{((2r-1)m+m_j+1/2)/m}(\Sigma). \quad \square \end{cases}$$

As for Proposition 11.1, Chapter 4, we verify that

$$(10.12)\qquad \mathscr{D}(\Sigma) \text{ is dense in } H^{\alpha m}\,\Xi^{\alpha}(\Sigma)$$

and therefore

$$H^{-\alpha m}\,\Xi^{-\alpha}(\Sigma) = (H^{\alpha m}\,\Xi^{\alpha}(\Sigma))' \subset \mathscr{D}'(\Sigma).$$

According to (10.11), we have:

$$(10.13)\quad \begin{cases} \text{if } \vec{g} \in \prod_{j=0}^{m-1} H^{-((2r-1)m+m_j+1/2)}\,\Xi^{-((2r-1)m+m_j+1/2)/m}(\Sigma), \\ \text{then } v \to \langle \vec{g}, \vec{T} v\rangle \text{ is continuous on } X^r(Q). \end{cases}$$

10.4 Choice of u_0 and u_1

$v \to \{v(0), v'(0)\}$ is a continuous mapping of

$$H^{(2r+1)m,(2r+1)}(Q) \to H^{(2r+1)m-m/2}(\Omega) \times H^{(2r+1)m-3m/2}(\Omega).$$

Using the spaces $\Xi^{\alpha}(\Omega)$, $\Xi^{-\alpha}(\Omega)$ introduced in Chapter 2, Section 6, it follows that:

$$(10.14)\quad \begin{cases} \text{if } \{u_0, u_1\} \in \Xi^{-(2r-1/2)m}(\Omega) \times \Xi^{-(2r+1/2)m}(\Omega), \\ \text{then the form } v \to \langle u_1, v(0)\rangle - \langle u_0, v'(0)\rangle \\ \text{is continuous on } X^r(Q). \quad \square \end{cases}$$

10.5 Conclusion

The preceding results may be summarized by

Theorem 10.1. *Let r be an integer, $r \geqq 1$. Take*

$$(10.15)\qquad f \in \Xi^{-(2r+1)m,-(2r+1)}(Q),$$

$$(10.16)\qquad \vec{g} \in \prod_{j=0}^{m-1} H^{-((2r-1)m+m_j+1/2)}\,\Xi^{-((2r-1)m+m_j+1/2)/m}(\Sigma),$$

$$(10.17)\qquad u_0 \in \Xi^{-(2r-1/2)m}(\Omega),$$

$$(10.18)\qquad u_1 \in \Xi^{-(2r+1/2)m}(\Omega);$$

then the solution u of (9.7) satisfies

$$(10.19)\qquad u \in D_{A+D_t^2}^{-(2r-1)}(Q) \quad \text{(see (10.7))}. \quad \square$$

It remains to specify (which we shall only do in a partial way) *in what sense* the solution u of (9.7) satisfies the relations

$$(10.20) \qquad A\,u + u'' = f, \quad \vec{B}\,u = \vec{g}, \quad u(0) = u_0, \quad u'(0) = u_1.$$

This is the objective of the next section.

11. Trace Theorems in the Space $D_{A+D_t^2}^{-(2r-1)}(Q)$

11.1 Density Theorem

Theorem 11.1. *Assume that hypotheses* (1.6), (1.8), (1.9) *are satisfied. Let u be a solution of* (9.7) *with the choices* (10.15), . . ., (10.18). *Then u belongs to the closure* $\mathbf{D}_{A+D_t^2}^{-(2r-1)}(Q)$ *of* $\mathscr{D}(\bar{Q})$ *in* $D_{A+D_t^2}^{-(2r-1)}(Q)$.

Before proving this theorem, we make

Remark 11.1. We do not know whether $\mathscr{D}(\bar{Q})$ is dense in $D_{A+D_t^2}^{-(2r-1)}(Q)$; see Problem 14.5 and Baiocchi [8], Hörmander [4, 5]. The application of the method of Theorem 10.1, Chapter 4, leads to a difficulty which we believe to be of sufficient interest to be discussed in some detail.

Let $u \to N(u)$ be a continuous antilinear form on $D_{A+D_t^2}^{-(2r-1)}(Q)$, such that $N(\varphi) = 0$, $\forall \varphi \in \mathscr{D}(\bar{Q})$. According to the Hahn-Banach theorem and definition (10.7), $N(u)$ may be represented by

$$(11.1) \qquad N(u) = \langle \psi_0, \bar{u} \rangle + \langle \psi_1, \overline{(A + D_t^2)\,u} \rangle,$$

where

$$\psi_0 \in H_{0,0}^{(2r-1)m,2r}(Q),$$

$$\psi_1 \in \Xi^{(2r+1)m,2r+1}(Q).$$

Let $\tilde{\psi}_0$, $\tilde{\psi}_1$ be the extensions of ψ_0 and ψ_1 to $\mathbf{R}_{x,t}^{n+1}$ by 0 outside Q. Let $\mathcal{O}$ be a bounded open set in $\mathbf{R}_x^n$, with regular boundary $\partial\mathcal{O}$, such that $\bar{\Omega} \subset \mathcal{O}$ and such that the operator A can be extended to an operator $\tilde{A}$ on

$$\mathcal{O} \times]-t_0, T + t_0[,$$

for a suitable $t_0 > 0$, which is elliptic in x, and has regular coefficients on $\mathcal{O} \times [-t_0, T + t_0]$.

For arbitrary Φ in $\mathscr{D}(\mathcal{O} \times]-t_0, T + t_0[)$, its restriction φ to Q belongs to $\mathscr{D}(\bar{Q})$, and therefore

$$(11.2) \qquad \langle \tilde{\psi}_0, \bar{\Phi} \rangle + \langle \tilde{\psi}_1, \overline{(\tilde{A} + D_t^2)\,\Phi} \rangle = N(\varphi) = 0,$$

where the brackets denote, *for example*, the duality between $\mathscr{D}'(\mathcal{O} \times]-t_0, T + t_0[)$ and $\mathscr{D}(\mathcal{O} \times]-t_0, T + t_0[)$. Therefore

$$(11.3) \qquad (\tilde{A}^* + D_t^2)\, \tilde{\psi}_1 = -\tilde{\psi}_0 \quad \text{in} \quad \mathcal{O} \times]-t_0, T + t_0[.$$

Note that, in particular, $\tilde{\psi}_1 \in L^2(\mathcal{O} \times] - t_0, T + t_0[)$.

Now let w be *the* solution of

$$(11.4) \qquad \begin{cases} (\tilde{A}^* + D_t^2)\, w = -\tilde{\psi}_0 \quad \text{in} \quad \mathcal{O} \times]-t_0, T + t_0[, \\ w(x, T + t_0) = 0, \quad w'(x, T + t_0) = 0, \\ \dfrac{\partial^j w}{\partial \nu^j} = 0, \quad 0 \leqq j \leqq m - 1, \quad \text{on} \quad \partial\mathcal{O} \times]-t_0, T + t_0[, \end{cases}$$

(*or* the same conditions "corresponding to B_j": $\tilde{B}_j w = 0$).

Since $\psi_0 \in H_{0,0}^{(2r-1)m, 2r}(Q)$, we have

$$\psi_0 \in H_{0,0}^{(2r-1)m, 2r}(\mathcal{O} \times]-t_0, T + t_0[)$$

and Theorem 7.1 yields

$$(11.5) \qquad w \in H^{(2r+1)m, 2r+1}(\mathcal{O} \times]-t_0, T + t_0[)$$

(a result which cannot be improved in an essential way, see Section 1).

Let us now verify that

$$(11.6) \qquad w = \tilde{\psi}_1.$$

To this end, let θ be arbitrary in $\mathscr{D}(\mathcal{O} \times]-t_0, T + t_0[)$ and let v be the solution of

$$(11.7) \qquad \begin{cases} (\tilde{A} + D_t^2)\, v = \theta \\ v(x, -t_0) = 0, \quad v'(x, -t_0) = 0, \\ \dfrac{\partial^j v}{\partial \nu^j} = 0 \quad \text{on} \quad \partial\mathcal{O} \times]-t_0, T + t_0[, \quad 0 \leqq j \leqq m - 1. \end{cases}$$

Since $\tilde{\psi}_1$ has compact support in $\mathcal{O} \times]-t_0, T + t_0[$, we have:

$$\langle (\tilde{A}^* + D_t^2)\, \tilde{\psi}_1, \bar{v} \rangle = \langle \tilde{\psi}_1, (\tilde{A} + D_t^2)\, \bar{v} \rangle = \langle \tilde{\psi}_1, \bar{\theta} \rangle,$$

and according to (11.3), we therefore have:

$$(11.8) \qquad \langle \tilde{\psi}_1, \bar{\theta} \rangle = - \langle \tilde{\psi}_0, \bar{v} \rangle.$$

On the other hand, Green's formula, valid "between" w and v, yields

$$\langle (\tilde{A}^* + D_t^2) w, \bar{v} \rangle = -\langle \tilde{\psi}_0, \bar{v} \rangle = \langle w, \overline{(\tilde{A} + D_t^2) v} \rangle = \langle w, \bar{\theta} \rangle,$$

which, by comparison with (11.8), shows (11.6).

But then, according to (11.5), $\tilde{\psi}_1$ belongs to

$$H^{(2r+1)m,2r+1}(\mathcal{O} \times]-t_0, T + t_0[)$$

and since $\tilde{\psi}_1$ vanishes outside Q we finally obtain

$$\psi_1 \in H_{0,0}^{(2r+1)m,2r+1}(Q), \tag{11.9}$$

and this regularity result cannot be improved in an essential way. But it *is not sufficient* to conclude that

$$\langle \psi_1, \overline{(A + D_t^2) u} \rangle = \langle (A^* + D_t^2) \psi_1, \bar{u} \rangle,$$

so that it does not follow that $N(u) = 0$, $\forall u$.

Here we see the *essential difference* between the present case and the parabolic, therefore hypoelliptic, case of Chapter 4, Theorem 10.1.

Of course, the preceding discussion does not constitute a counter-example! □

Remark 11.2 (due to C. Baiocchi). If we denote by $D^{\infty}_{A+D_t^2}(Q)$ the space of u's belonging to $\mathscr{D}'(Q)$ such that

$$A u + u'' \in \Xi^{-(2r+1)m,-(2r+1)}(Q),$$

then the preceding arguments, this time, yield the fact that $\mathscr{D}(\bar{Q})$ is dense in $D^{\infty}_{A+D_t^2}(Q)$. □

Proof of Theorem 11.1. Since $\mathscr{D}(Q)$ (resp. $\mathscr{D}(\Sigma)^m$, resp. $\mathscr{D}(\Omega)$, resp. $\mathscr{D}(\Omega)$) is dense in

$$\Xi^{-(2r+1)m,-(2r+1)}(Q)$$

$$\left(\text{resp.} \prod_{j=0}^{m-1} H^{-((2r-1)m+m_j+1/2)} \Xi^{-((2r-1)m+m_j+1/2)/m}(\Sigma),\right.$$

$$\left.\text{resp.}\ \Xi^{-(2r-1/2)m}(\Omega),\ \text{resp.}\ \Xi^{-(2r+1/2)m}(\Omega)\right),$$

we can find

$$f_n \in \mathscr{D}(Q), \quad \vec{g}_n \in \mathscr{D}(\Sigma)^m, \quad u_{0n} \in \mathscr{D}(\Omega), \quad u_{1n} \in \mathscr{D}(\Omega), \tag{11.10}$$

with

$$f_n \to f \quad \text{in} \quad \Xi^{-(2r+1)m,-(2r+1)}(Q), \tag{11.11}$$

$$\vec{g}_n \to \vec{g} \quad \text{in} \quad \prod_{j=0}^{m-1} H^{-((2r-1)m+m_j+1/2)} \Xi^{-((2r-1)m+m_j+1/2)/m}(\Sigma), \tag{11.12}$$

$$u_{0n} \to u_0 \quad \text{in} \quad \Xi^{-(2r-1/2)m}(\Omega), \tag{11.13}$$

$$u_{1n} \to u_1 \quad \text{in} \quad \Xi^{-(2r+1/2)m}(\Omega). \tag{11.14}$$

Set (compare with (9.8)):

$$(11.15) \qquad L_n(v) = \langle f_n, \bar{v}\rangle + \langle \vec{g}_n, \overline{\vec{T} v}\rangle + \langle u_{1n}, \overline{v(0)}\rangle - \langle u_{0n}, \overline{v'(0)}\rangle .$$

Let (applying Theorem 9.1) u_n be *the* solution in $H^{-(2r-1)m,-2r}(Q)$ of

$$(11.16) \qquad (u_n, A^* v + v'') = L_n(v), \quad \forall v \in X^r(Q).$$

As $n \to \infty$, $L_n(v) \to L(v)$ uniformly for v in a bounded set of $(X^r Q)$ and therefore

$$(11.17) \qquad u_n \to u \quad \text{in} \quad H^{-(2r-1)m,-2r}(Q).$$

Since, on the other hand:

$$A u_n + u_n'' = f_n,$$

we have, according to (11.11):

$$(11.18) \qquad A u_n + u_n'' \to A u + u'' \quad \text{in} \quad \Xi^{-(2r+1)m,-(2r+1)}(Q)$$

and from (11.17)–(11.18) follows that

$$(11.19) \qquad u_n \to u \quad \text{in} \quad D_{A+D_t^2}^{-(2r-1)}(Q).$$

But let us a priori consider $\tilde{u}_n$ as a solution of

$$(11.20) \qquad \begin{cases} (A + D_t^2)\, \tilde{u}_n = f_n, \\ B\, \tilde{u}_n = \vec{g}_n, \\ \tilde{u}_n(x,0) = u_{0n}, \quad \dfrac{\partial \tilde{u}_n}{\partial t}(x,0) = u_{1n}. \end{cases}$$

Then, according to the *regularity Theorem* 7.1 (the problem can be easily reduced to the "homogeneous" case: $\vec{g}_n = 0$, $u_{0n} = 0$, $u_{1n} = 0$) it follows that

$$(11.21) \qquad \tilde{u}_n \in \mathscr{D}(\bar{Q}).$$

But then the *usual* Green's formula yields

$$\langle \tilde{u}_n, \overline{(A^* + D_t^2)\, v}\rangle = L_n(v), \quad \forall v \in X^r(Q)$$

and since (11.16) admits a *unique* solution, we have

$$(11.22) \qquad \tilde{u}_n = u_n .$$

The desired result follows from (11.21), (11.22) and (11.19).

11.2 Traces on Σ

Theorem 11.2. *Let r be an integer, $r \geqq 1$ and*

$$2r + 2 - \left(\frac{2r+2}{2r+1}\right)\left(\frac{k+\frac{1}{2}}{m}\right) \neq \textit{integer} + \tfrac{1}{2},$$

$$\textit{for} \quad 0 \leqq k \leqq (2r+1)\,m - 1.$$

The mapping $u \to B_j u$ of $\mathscr{D}(\bar{Q})$ into $\mathscr{D}(\Sigma)$ extends by continuity to a continuous linear mapping, still denoted by $u \to B_j u$, of $\mathbf{D}_{A+D_t^2}^{-(2r-1)}(Q)$ (see Theorem 11.1) $\to H^{-(2r+1)m+2m-m_j-1/2,(2r+2)+\left(\frac{2r+2}{2r+1}\right)\left(\frac{2m-m_j-1/2}{m}\right)}(\Sigma)$.

Proof. Let us, for the time being, admit

Lemma 11.1. *Let r be an integer $\geqq 1$ with*

$$2r + 2 - \left(\frac{2r+2}{2r+1}\right)\left(\frac{k+\frac{1}{2}}{m}\right) \neq \textit{integer} + \tfrac{1}{2},$$

$$\textit{for} \quad 0 \leqq k \leqq (2r+1)\,m - 1;$$

then there exists a right inverse

$$\vec{g} = \{g_j\}_{j=0}^{m-1} \to R(\vec{g}) = v, \tag{11.23}$$

which is a continuous linear mapping of

$$\prod_{j=0}^{m-1} H_{\;\;,0}^{(2r+1)m-(2m-m_j-1/2),2r+2-\left(\frac{2r+2}{2r+1}\right)\left(\frac{2m-m_j-1/2}{m}\right)}(\Sigma) \to \tag{11.24}$$
$$\to H_{\;\;,0}^{(2r+1)m,2r+2}(Q)$$

such that

$$(A^* + D_t^2)\, v \in H_{0,0}^{(2r-1)m,2r}(Q), \tag{11.25}$$

$$\vec{C}\, v = 0, \tag{11.26}$$

$$\vec{T}\, v = \vec{g}. \quad \square \tag{11.27}$$

For given u in $\mathbf{D}_{A+D_t^2}^{-(2r-1)}(Q)$, set

$$Z(\vec{g}) = -\langle A\,u + D_t^2 u, \bar{v}\rangle + \langle u, \overline{A^*v} + \overline{D_t^2 v}\rangle, \tag{11.28}$$

where the first (resp. second) bracket denotes the duality between

$\Xi^{-(2r+1)m,-(2r+1)}(Q)$ and $\Xi^{(2r+1)m,2r+1}(Q) \supset H^{(2r+1)m,2r+1}(Q) \supset$

$\supset H_{\;\;,0}^{(2r+1)m,2r+2}(Q)$ (resp. $H^{-(2r-1)m,-2r}(Q)$ and $H_{0,0}^{(2r-1)m,2r}(Q)$).

We verify that $Z(\vec{g})$ depends only on $\vec{g}$ and not on the choice of the (non-unique) right inverse satisfying the conditions of Lemma 11.1. Indeed, if v_1 is a second function satisfying these conditions, then $v - v_1 = w$ satisfies (since $\{\vec{C}, \vec{T}\}$ is a Dirichlet system; Chapter 2, Section 1)

$$\frac{\partial^k w}{\partial \nu^k} = 0 \quad \text{on} \quad \Sigma, \qquad 0 \leqq k \leqq 2m-1.$$

Next, $A^* w + D_t^2 w \in H_{0,0}^{(2r-1)m, 2r}(Q)$ implies (as in Theorem 10.2 of Chapter 4) that $\dfrac{\partial^k w}{\partial \nu^k} = 0$ on Σ for $0 \leqq k \leqq (2r+1)m - 1$, and finally $w \in H_{0,0}^{(2r+1)m, 2r+2}(Q)$, from which the result follows.

The antilinear form $\vec{g} \to Z(\vec{g})$ is continuous in $\vec{g}$, therefore

$$Z(\vec{g}) = \langle \vec{\beta}_u, \vec{g} \rangle, \tag{11.29}$$

where

$$\begin{cases} \vec{\beta}_u \in \prod\limits_{j=0}^{m-1} \left(H^{(2r+1)m-(2m-m_j-1/2), 2r+2-\left(\frac{2r+2}{2r+1}\right)\left(\frac{2m-m_j-1/2}{m}\right)}(\Sigma) \right)' \\ = \prod\limits_{j=0}^{m-1} H^{-(2r+1)m+2m-m_j-1/2, -(2r+2)+\left(\frac{2r+2}{2r+1}\right)\left(\frac{2m-m_j-1/2}{m}\right)}(\Sigma) \\ = \mathscr{T}. \end{cases} \tag{11.30}$$

In this way, we have defined a linear mapping

$$u \to \vec{\beta}_u \quad \text{of} \quad \mathbf{D}_{A+D_t^2}^{-(2r-1)}(Q) \to \mathscr{T}$$

which is *continuous* (as in Theorem 10.2, Chapter 4).

Finally, we verify (starting from Green's formula) that

$$\vec{\beta}_u = \vec{B} u \quad \text{if} \quad u \in \mathscr{D}(\bar{Q}),$$

which, thanks to Theorem 11.1, proves the theorem, if we can prove Lemma 11.1. □

Proof of Lemma 11.1. We use the same principle as for the proof of Lemma 10.1, Chapter 4. Conditions (11.26)–(11.27) are equivalent to the data

$$\begin{cases} \left.\dfrac{\partial^k v}{\partial \nu^k}\right|_{\Sigma} = h_k, \quad 0 \leqq k \leqq 2m-1, \\ h_k \in H_{\ ,0}^{(2r+1)m-(k+1/2), (2r+2)-\left(\frac{2r+2}{2r+1}\right)\left(\frac{k+1/2}{m}\right)}(\Sigma). \end{cases} \tag{11.31}$$

Next, we write the boundary conditions corresponding to (11.24) and (11.25):

$$\frac{\partial^k A^* v}{\partial \nu^k} + \frac{\partial^k v''}{\partial \nu^k} = 0 \quad \text{on} \quad \Sigma, \quad 0 \leqq k \leqq (2r-1)\, m - 1, \tag{11.32}$$

$$\frac{\partial^j v(x,0)}{\partial t^j} = \frac{\partial^j v(x,T)}{\partial t^j} = 0, \quad x \in \Omega, \quad 0 \leqq j \leqq 2r + 1. \tag{11.33}$$

We apply (11.32) successively: first for $0 \leqq k \leqq 2m - 1$, we may use (11.31) and therefore

$$\frac{\partial^k A^* v}{\partial \nu^k} = -h_k'' \in H^{(2r+1)m-(k+1/2),\, 2r - \left(\frac{2r+2}{2r+1}\right)\left(\frac{k+1/2}{m}\right)}_{,0} (\Sigma)$$

therefore, *in particular:*

$$h_k'' \in H^{(2r+1)m-(k+1/2),\, (2r+2) - \left(\frac{2r+1}{2r+1}\right)\left(\frac{k+2m+1/2}{m}\right)}_{,0} (\Sigma)$$

and finally it follows that the boundary conditions are equivalent to

$$\left\{ \begin{aligned} & \frac{\partial^k v}{\partial \nu^k} = h_k \in H^{(2r+1)m-(k+1/2),\, (2r+2) - \left(\frac{2r+2}{2r+1}\right)\left(\frac{k+1/2}{m}\right)}_{,0} (\Sigma) \\ & \text{for} \quad 0 \leqq k \leqq (2r+1)\, m - 1. \end{aligned} \right. \tag{11.34}$$

The lemma then follows from Theorem 2.3, Chapter 4[1]. □

Remark 11.3. If

$$\nu_j = 2r + 2 - \left(\frac{2r+2}{2r+1}\right)\left(\frac{j+\frac{1}{2}}{m}\right)$$

equals an integer $+ \frac{1}{2}$, then Lemma 11.1 contains integral compatibility relations. Indeed if $u \in H^{(2r+1)m,\, 2r+2}_{,0}(Q)$, then

$$\frac{\partial^j u}{\partial \nu^j} = \varphi_j, \quad \frac{\partial^k u}{\partial t^k} = 0$$

and, assuming that Ω is the half-space $x_n > 0$, we must have (see Theorem 2.3, Chapter 4)

$$\int_0^\infty \int_{R^{n-1}} \left| \frac{\partial^k}{\partial t^k} \varphi_j(x', \sigma^{(2r+1)m}) \right|^2 dx' \frac{d\sigma}{\sigma} < \infty$$

[1] Applied with $f_k = 0$ and $g_j = h_j$.

if

$$\frac{j}{(2r+1)m} + \frac{k}{2r+2} = 1 - \frac{1}{2}\left(\frac{1}{(2r+1)m} + \frac{1}{2r+2}\right)$$

which is equivalent to $\nu_j = k + \frac{1}{2}$.

Of course, this corresponds to complicated integral relations on the g_j's in Lemma 11.1. □

11.3 Continuity of the Trace on Neighbouring Surfaces

The notation is the same as in Chapter 4.

For $u \in \tilde{\mathbf{D}}_{A+D_t^2}^{-(2r-1)}(Q)$ (= closure of $\mathscr{D}(\bar{Q})$ in the space $\tilde{D}_{A+D_t^2}^{-(2r-1)}(Q)$ defined by $\{u \mid u \in H^{-(2r-1)m,-2r}(Q),\ A\,u + D_t^2\,u \in L^2(Q)\}$), we define

$$(11.35)\quad B_j^{(\varrho)}\,u \in H^{-(2r+1)m+2m-m_j-1/2,\,-(2r+2)+\left(\frac{2r+2}{2r+1}\right)\left(\frac{2m-m_j-1/2}{m}\right)}(\Sigma)$$

("trace of $B_j\,u$" on "neighbouring surfaces" of Σ).

By the same method as for Theorem 10.3, Chapter 4, we show

Theorem 11.3. *Under the hypotheses of Theorem 11.2, if $u \in \tilde{\mathbf{D}}_{A+D_t^2}^{-(2r-1)}(Q)$, we have*

$$(11.36)\quad \begin{cases} B_j^{(\varrho)}\,u \to B_j\,u \quad \textit{in} \\ H^{-(2r+1)m+2m-m_j-1/2,\,-(2r+2)+\left(\frac{2r+2}{2r+1}\right)\left(\frac{2m-m_j-1/2}{m}\right)}(\Sigma), \\ \textit{as}\quad \varrho \to 0. \quad \square \end{cases}$$

11.4 Traces on Ω_0

Theorem 11.4. *Let r be an integer, $r \geqq 1$, with*

$$(2r+1)\,m - \left(\frac{2r+1}{2r+2}\right)\left(\frac{i+1}{2}\right) \neq \textit{integer} + \frac{1}{2}, \qquad i = 0, 1^{((1))}.$$

The mapping

$$u \to \{u(x,0),\, u'(x,0)\} \quad \textit{of} \quad \mathscr{D}(\bar{Q}) \to \mathscr{D}(\bar{\Omega}) \times \mathscr{D}(\bar{\Omega})$$

extends by continuity to a continuous linear mapping, still denoted by

$$u \to \{u(x,0),\, u'(x,0)\},$$

of

$$\mathbf{D}_{A+D_t^2}^{-(2r-1)}(Q) \to H^{-(2r+1)m+3/2\left(\frac{2r+1}{2r+2}\right)}(\Omega) \times H^{-(2r+1)m+1/2\left(\frac{2r+1}{2r+2}\right)}(\Omega).$$

((1)) For the exceptional cases, a remark similar to Remark 11.3 holds.

Proof. As for Lemma 11.1, we verify that there exists a right inverse

$$\{\varphi_0, \varphi_1\} = \vec{\varphi} \to R_1(\vec{\varphi}) = v_1,$$

which is a continuous linear mapping of

$$H_0^{(2r+1)m-\left(\frac{2r+1}{2r+2}\right)\frac{1}{2}}(\Omega) \times H_0^{(2r+1)m-\left(\frac{2r+1}{2r+2}\right)\frac{3}{2}}(\Omega) \to H_{0,}^{(2r+1)m,2r+2}(Q)$$

such that

$$v_1(x,0) = \varphi_0, \quad v_1'(x,0) = \varphi_1,$$
$$A^* v_1 + D_t^2 v_1 \in H_{0,0}^{(2r-1)m,2r}(Q),$$
$$v_1(x,T) = 0, \quad v_1'(x,T) = 0.$$

For u given in $\mathbf{D}_{A+D_t^2}^{-(2r-1)}(Q)$, we define:

(11.37) $$Z_1(\vec{\varphi}) = -\langle A\,u + u'', \bar{v}_1\rangle + \langle u, \overline{A^* v_1 + v_1''}\rangle,$$

the duality being taken as in (11.28).

We verify that $Z_1(\vec{\varphi})$ is independent of the choice of the right inverse and consequently that

$$Z_1(\vec{\varphi}) = \langle \tau_u, \{\varphi_0, \varphi_1\}\rangle$$

$$\tau_u \in H^{-(2r+1)m+\left(\frac{2r+1}{2r+2}\right)\frac{1}{2}}(\Omega) \times H^{-(2r+1)m+\left(\frac{2r+1}{2r+2}\right)\frac{3}{2}}(\Omega),$$

the mapping $u \to \tau_u$ being linear and continuous.

Next, we verify that if $\tau_u = \{\tau_u^0, \tau_u^1\}$,

$$Z_1(\vec{\varphi}) = \langle \tau_u^0, \bar{\varphi}_0\rangle + \langle \tau_u^1, \bar{\varphi}\rangle,$$

then $\tau_u^0 = -u'(x,0)$, $\tau_u^1 = u(x,0)$ if $u \in \mathcal{D}(\bar{Q})$. Whence the theorem.

11.5 Continuity of the Trace on Sections Neighbouring Ω_0

In the notation of Chapter 4, we have:

Theorem 11.5. *Assume* (1.6), (1.8) *and* (1.9) *to be satisfied. Let integer* $r \geq 1$ *with*

$$(2r+1)\,m - \left(\frac{2r+1}{2r+2}\right)\left(\frac{i+1}{2}\right) \neq \textit{integer} + \frac{1}{2}, \quad i = 0, 1.$$

For u *given in* $\tilde{\mathbf{D}}_{A+D_t^2}^{-(2r-1)}(Q)$, *we have:*

(11.38) $$u(x,t_0) \to u(x,0) \quad \textit{in} \quad H^{-(2r+1)m+\left(\frac{2r+1}{2r+2}\right)\frac{3}{2}}(\Omega),$$

(11.39) $$u'(x,t_0) \to u'(x,0) \quad \textit{in} \quad H^{-(2r+1)m+\left(\frac{2r+1}{2r+2}\right)\frac{1}{2}}(\Omega),$$

as $t_0 \to 0$. □

11.6 Remark

The spaces appearing in (10.16), (10.17) and (10.18) are contained in the spaces appearing in the trace Theorems 11.2 and 11.4. Theorems 11.2 and 11.4 give meaning to (10.20). But we must be careful of the fact that we do not have "optimal" results. Formulation (10.20) is only a weakened (but intuitive) formulation of Theorem 9.1. As we have done for the parabolic case in Section 12.3, Chapter 4, we could seek a *characterization* of the traces of u's belonging to $\mathbf{D}_{A+D_t^2}^{-(2r-1)}(Q)$, but we would find the same type of difficulties. Indeed, we can *globally*, by extension by continuity, define the operator $u \to \tau u = \{\vec{B}u, u(x,0), D_t u(x,0)\}$ for all $u \in \mathbf{D}_{A+D_t^2}^{-(2r-1)}(Q)$ and *characterize* the space $\mathscr{V}'_r$ described by τu as u describes $\mathbf{D}_{A+D_t^2}^{-(2r-1)}(Q)$, so that the problem

$$A\,u + D_t^2 u = f; \qquad \tau u = g^*$$

admits a unique solution $u \in \mathbf{D}_{A+D_t^2}^{-(2r-1)}(Q)$ for all $f \in \Xi^{-(2r+1)m,-(2r+1)}(Q)$ and $g^* \in \mathscr{V}'_r$. But then we meet the same difficulties as in Section 12.3, Chapter 4, if we want to separate the operators $\vec{B}\,u$, $u(x, 0)$ and $D_t u(x,0)$ in τu. We shall not insist on this point (see Problem 14.1). □

12. Schroedinger Type Equations

12.1 Notation

The operator A and the boundary operators B_j have the *same* properties as in Section 1. For Theorem 12.2, we shall also require the hypothesis

$$A^{(j)}(t) \in \mathscr{L}(H^0(\Omega); H^0(\Omega)), \quad \forall j \geqq 1 \tag{12.1}$$

(that is, the functions $a_{pq}(x,t)$ *are independent of* t, except $a_{00}(x,t)$).

We shall consider the (non-homogeneous Schroedinger type) problems:

$$\begin{cases} \mathrm{i}\,A\,u + u' = f, \\ \vec{B}\,u = \vec{g} \\ u(x,0) = u_0. \quad \square \end{cases} \tag{12.2}$$

We shall give two regularity theorems. And we shall introduce the so-called method of *parabolic regularization* (see also Chapter 3, Section 8.5).

12.2 First Regularity Theorem. Parabolic Regularization

Theorem 12.1. *Assume the hypotheses of Section 1 to be satisfied. Let f be given with*

$$f, f' \in L^2(Q) \qquad (\text{i.e. } f \in H^{0,1}(Q)) \tag{12.3}$$

and u_0 be given with

$$u_0 \in H^{2m}(\Omega), \quad B_j\left(x, 0, \frac{\partial}{\partial x}\right) u_0 = 0, \quad 0 \leqq j \leqq m-1. \tag{12.4}$$

Then the solution u of problem (12.2), *with $\vec{g} = 0$, satisfies*

$$u \in H^{2m,1}(Q) \tag{12.5}$$

(i.e.: $u \in L^2(0, T; H^{2m}(\Omega))$, $u' \in L^2(0, T; H^0(\Omega))$).

Proof.

1) *Variational formulation.*

In the notations of Section 1 and Chapter 3, problem (12.2) is equivalent to:

$$\begin{cases} \mathrm{i}\, a(t; u(t), v) + (u'(t), v) = (f(t), v) & \forall v \in V, \\ u \in L^2(0, T; V), \quad u(0) = u_0. \end{cases} \tag{12.6}$$

2) *Parabolic regularization* (see also Chapter 3, Section 8.5).

Let $\varepsilon > 0$ be arbitrary. Denote by u_ε *the* solution of the problem (which we shall call the *parabolic regularization* of (12.5)):

$$(\varepsilon + \mathrm{i})\, a(t; u_\varepsilon, v) + (u'_\varepsilon, v) = (f, v), \quad \forall v \in V. \tag{12.7}$$

Letting $v = u_\varepsilon$ in (12.7) and taking twice the real part of the result, we obtain:

$$2\varepsilon\, a(t; u_\varepsilon, u_\varepsilon) + \frac{d}{dt} |u_\varepsilon(t)|^2 = (f(t), u_\varepsilon(t)),$$

and therefore

$$\frac{d}{dt} |u_\varepsilon(t)|^2 \leqq |f(t)|\, |u_\varepsilon(t)|$$

and consequently

(12.8) u_ε remains in a bounded set of $L^2(0, T; H^0(\Omega))$, as $\varepsilon \to 0$.

3) According to Chapter 4, we know that we can differentiate (12.7) with respect to t:

$$u'_\varepsilon \in L^2(0, T; V)$$

satisfies

$$(12.9)\quad \begin{cases} (\varepsilon+\mathrm{i})\, a(t; u_\varepsilon', v) + (\varepsilon+\mathrm{i})\, a'(t; u_\varepsilon, v) + (u_\varepsilon'', v) = (f', v) \quad \forall v \in V, \\ u_\varepsilon(0) = u_0, \quad u_\varepsilon'(0) = f(0) - (\varepsilon+\mathrm{i}) A(0) u_0 = u_{1\varepsilon} \in H^0(\Omega). \end{cases}$$

If we replace v with $u_\varepsilon(t)$ in (12.9) and take the real part, we obtain

$$(12.10)\quad \begin{cases} \varepsilon\, a(t; u_\varepsilon'(t), u_\varepsilon'(t)) + \dfrac{1}{2} \dfrac{d}{dt} |u_\varepsilon'(t)|^2 \\ = \mathrm{Re}(f', u_\varepsilon'(t)) - \mathrm{Re}(\varepsilon + \mathrm{i})\, a'(t; u_\varepsilon, u_\varepsilon'). \end{cases}$$

But, thanks to (12.1):

$$|a'(t; u_\varepsilon, u_\varepsilon')| \leqq c\, |u_\varepsilon(t)|\, |u_\varepsilon'(t)|$$

(where $|\ |$ denotes the norm in $H^0(\Omega)$) and thus it follows from (12.8) and (12.10) that

(12.11) $\begin{cases} u_\varepsilon' \textit{ remains} \text{ (in particular) } \textit{in a bounded set} \\ \textit{of } L^2(0, T; H^0(\Omega)),\ \textit{as } \varepsilon \to 0. \end{cases}$

On the other hand $\sqrt{\varepsilon}\, u_\varepsilon'$ remains in a bounded set of $L^2(0, T; V)$.

4) Letting $v = u_\varepsilon$ in (12.7) and, this time, taking the *imaginary parts*, we obtain:

$$a(t; u_\varepsilon(t), u_\varepsilon(t)) = \mathrm{Im}(f, u_\varepsilon(t)) - \mathrm{Im}(u_\varepsilon', u_\varepsilon)$$

and thanks to (12.11) it follows that

(12.12) u_ε remains in a bounded set of $L^2(0, T; V)$, as $\varepsilon \to 0$.

5) We can therefore extract a sequence — still denoted by u_ε — from u_ε, such that

$$u_\varepsilon \to u \quad \text{weakly in} \quad L^2(0, T; V),$$
$$u_\varepsilon' \to u \quad \text{weakly in} \quad L^2(0, T; H), \qquad H = H^0(\Omega),$$

and necessarily u = solution of (12.6). By uniqueness, it is unnecessary to extract a "subsequence" from u_ε.

Thus we have shown the (weak; see Remark 12.1) convergence of u_ε to u and that the solution u of (12.6) satisfies

$$u \in L^2(0, T; V), \qquad u' \in L^2(0, T; H).$$

But

$$\begin{cases} \mathrm{i}\, a(t; u(t), v) = (f(t) - u'(t), v), \qquad \forall v \in V, \\ f - u' \in L^2(0, T; H) \end{cases}$$

implies (*elliptic* regularization) that

$$u \in L^2(0, T; H^{2m}(\Omega))$$

and this completes the proof of the theorem. □

Remark 12.1. We have:

$$\mathrm{i}\, a(t; u - u_\varepsilon, u - u_\varepsilon) + (u' - u'_\varepsilon, u - u_\varepsilon) = \varepsilon\, a(t; u_\varepsilon, u - u_\varepsilon)$$

and hence

$$\frac{d}{dt} |u(t) - u_\varepsilon(t)|^2 = 2\varepsilon \operatorname{Re} a(t; u_\varepsilon, u - u_\varepsilon)$$

and therefore *in particular* $u_\varepsilon \to u$ *strongly* in $L^2(0, T; H^0(\Omega))$, as $\varepsilon \to 0$. □

12.3 Second Regularity Theorem

Theorem 12.2. *Assume the hypotheses of Section* 1 *and* (12.1) *to be satisfied. Let f be given with*

(12.13) $$f \in H^{(2r-1)m, r}_{\,,0}(Q), \quad \textit{integer } r \geqq 1.$$

Let

(12.14) $$\vec{g} = 0, \quad u_0 = 0.$$

Then the solution u of problem (12.2) *satisfies*

(12.15) $$u \in H^{2rm, r}(Q).$$

Proof. The theorem is valid for $r = 1$ (consequence of Theorem 12.1). We assume the theorem to hold up to $r - 1$, that is

$$u \in H^{2(r-1)m, r-1}(Q),$$

and we show (12.15).

Again, let u_ε be the *parabolic regularization* of u (see Section 12.2), that is the solution of

$$(\varepsilon + \mathrm{i})\, a(t; u_\varepsilon, v) + (u'_\varepsilon, v) = (f, v).$$

Differentiating r times with respect to t:

(12.16) $$(\varepsilon + \mathrm{i})\, a(t; u_\varepsilon^{(r)}, v) + (\varepsilon + \mathrm{i}) \sum_{j=0}^{r-1} \binom{r}{j} a^{(r-j)}(t; u_\varepsilon^{(j)}, v) + \\ + (u_\varepsilon^{(r+1)}, v) = (f^{(r)}, v); \quad u_\varepsilon^{(j)}(0) = 0, \quad 0 \leqq j \leqq r.$$

Replacing v with $u_\varepsilon^{(r)}(t)$ in (12.16) and taking the real part, we obtain

$$(12.17)\quad \begin{cases} \varepsilon\, a(t; u_\varepsilon^{(r)}(t), u_\varepsilon^{(r)}(t)) + \dfrac{1}{2}\dfrac{d}{dt}|u_\varepsilon^{(r)}(t)|^2 \\ \qquad = \operatorname{Re}(f_\varepsilon^{(r)}, u^{(r)}) - \operatorname{Re}(\varepsilon + \mathrm{i}) \displaystyle\sum_{j=0}^{r-1} \binom{r}{j} a^{(r-j)}(t; u_\varepsilon^{(j)}, u_\varepsilon^{(r)}). \end{cases}$$

It follows that

$$(12.18)\quad \begin{cases} \sqrt{\varepsilon}\, u_\varepsilon^{(r)} \text{ remains in a bounded set of } L^2(0, T; V), \\ \text{as } \varepsilon \to 0 \text{ and } u_\varepsilon^{(r)} \text{ remains in a bounded set of} \\ L^2(0, T; H^0(\Omega)). \end{cases}$$

The derivative of order $(r-1)$, instead of (12.16), and after having replaced v with $u_\varepsilon^{(r-1)}$, leads to

$$(12.19)\quad \begin{cases} (\varepsilon + \mathrm{i})\, a(t; u_\varepsilon^{(r-1)}, u_\varepsilon^{(r-1)}) + \\ \qquad + (\varepsilon + \mathrm{i}) \displaystyle\sum_{j=0}^{r-2} \binom{r-1}{j} a^{(r-1-j)}(t; u_\varepsilon^{(j)}, u_\varepsilon^{(r-1)}) + \\ \qquad + (u_\varepsilon^{(r)}, u_\varepsilon^{(r-1)}) = (f^{(r-1)}, u_\varepsilon^{(r-1)}), \end{cases}$$

and therefore, taking the *imaginary* part of (12.19) and using (12.18), it follows that

$$(12.20)\quad u_\varepsilon^{(r-1)} \text{ remains in a bounded set of } L^2(0, T; V), \text{ as } \varepsilon \to 0.$$

Extracting a suitable sequence and letting $\varepsilon \to 0$, it follows that

$$(12.21)\quad \begin{cases} u_\varepsilon^{(r)} \in L^2(0, T; H^0(\Omega)) = L^2(Q), \\ u^{(r-1)} \in L^2(0, T; H^m(\Omega)), \end{cases}$$

$$(12.22)\quad \begin{cases} \mathrm{i}\, a(t; u^{(r-1)}, v) + \mathrm{i} \displaystyle\sum_{j=0}^{r-2} \binom{r-1}{j} a^{(r-1-j)}(t; u^{(j)}, v) + \\ \qquad + (u^{(r)}, v) = (f^{(r-1)}, v) \quad \forall v \in V. \end{cases}$$

But according to the induction hypothesis and the intermediate derivatives theorem (Chapter 1, Section 3), we already know that

$$u^{(j)} \in L^2(0, T; H^{2(r-1)m(1-j/(r-1))}(\Omega))$$

and therefore

$$A^{(r-1-j)} u^{(j)} \in L^2(0, T; H^{2m(r-1-j-1)}(\Omega)) \subset L^2(0, T; H^0(\Omega)),$$

since

$$j \leqq r - 2.$$

Therefore it follows from (12.22) (the $u^{(j)}$'s satisfying $B_k u^{(j)} = 0$) that

$$\mathrm{i}\, A(t)\, u^{(r-1)}(t) = f^{(r-1)} - u^{(r)} - \mathrm{i} \sum_{j=0}^{r-2} \binom{r-1}{j} \times$$

$$\times A^{(r-j-1)} u^{(j)} \in L^2(0, T; H^0(\Omega))$$

and since $B_k u^{(r-1)} = 0$, $0 \leqq k \leqq m - 1$, it follows that

(12.23) $$u^{(r-1)} \in L^2(0, T; H^{2m}(\Omega)).$$

But, differentiating the equation

$$\mathrm{i}\, a(t; u, v) + (u', v) = (f, v)$$

$(r - 2)$ times with respect to t (which is now permissible), we obtain

(12.24) $$\mathrm{i}\, a(t; u^{(r-2)}, v) + \mathrm{i} \sum_{j=0}^{r-3} \binom{r-2}{j} a^{(r-2-j)}(t; u^{(j)}, v) +$$

$$+ (u^{(r-1)}, v) = (f^{(r-2)}, v) \quad \forall v \in V.$$

Now, we have (12.23). Furthermore

$$A^{(r-2-j)} u^{(j)} \in L^2(0, T; H^{2m}(\Omega))$$

and (12.13), together with the intermediate derivatives theorem, yields

$$f^{(r-2)} \in L^2(0, T; H^{2(r-1)m(1-(r-2)/r)}(\Omega)) \subset L^2(0, T; H^{2m}(\Omega))$$

(we see that here we could *weaken the hypotheses* on f) and (12.24) yields

$$u^{(r-2)} \in L^2(0, T; H^{4m}(\Omega)).$$

And so on, taking the derivative with respect to t, of order $r - 3$, $r - 4, \ldots$ □

Remark 12.2. As in Remark 7.3, we see that Theorem 12.2 is essentially the best possible result. In the notation of Remark 7.3, we have:

$$\hat{u} = G \underset{(t)}{*} f,$$

where

$$G = G(\lambda, t) = \begin{cases} \exp(-\mathrm{i}\,\lambda\, t) & \text{if } t > 0 \\ 0 & \text{if } t < 0. \end{cases}$$

From the formula

$$G = \frac{1}{\mathrm{i}\,\lambda}\left[\delta + \frac{1}{(-\mathrm{i}\,\lambda)}\delta' + \frac{1}{(-\mathrm{i}\,\lambda)^2}\delta'' + \cdots + \frac{1}{(-\mathrm{i}\,\lambda)^{r-1}}\delta^{(r-1)} + \right.$$
$$\left. + \frac{1}{(-\mathrm{i}\,\lambda)^{r-1}} G^{(r)}\right]$$

$\left(\text{where } G^{(r)} = \dfrac{d^r}{dt^r} G\right)$, we deduce that

$$\hat{u} = \frac{1}{\mathrm{i}\,\lambda}\hat{f} + \frac{1}{\mathrm{i}\,\lambda(-\mathrm{i}\,\lambda)}\hat{f}' + \cdots + \frac{1}{(\mathrm{i}\,\lambda)\,(-\mathrm{i}\,\lambda)^{r-1}}\hat{f}^{(r-1)} +$$
$$+ \frac{1}{\mathrm{i}\,\lambda(-\mathrm{i}\,\lambda)^{r-1}} G * \hat{f}^{(r)}.$$

Therefore if we assume slightly more than (12.13) (see Remark 7.3), that is:

$$\lambda^{r-1}\hat{f} \in L^2(0, T; \mathfrak{h}) \quad \text{and} \quad \hat{f}^{(r)} \in L^2(0, T; \mathfrak{h}),$$

then we verify that

$$\lambda^r\, \hat{u} \in L^2(0, T; \mathfrak{h})$$

and with no possibility of improvement (in the Hilbert setting). □

12.4 *r*-Isomorphism Theorem

We shall give only the *steps* corresponding to Sections 6 to 10. (The point of view corresponding to Sections 1 to 5 is not specified here.)

We consider the *adjoint problem*:

$$\text{(12.25)} \qquad \begin{cases} -\mathrm{i}\, A^* v - v' = \varphi, \\ C\, v = 0, \\ v(x, T) = 0, \end{cases}$$

and we assume that

$$\text{(12.26)} \qquad \varphi \in H^{2(r-1)m, r}_{0,0}(Q),$$

integer $r \geqq 1$ (*by interpolation*, we could also consider the case "non-integer r").

Define

$$(12.27)\quad \begin{cases} X^r(Q) = \text{space described by the solution } v \text{ of (12.25)} \\ \qquad \text{as } \varphi \text{ describes } H^{2(r-1)m,r}_{0,0}(Q), \text{ provided with the} \\ \qquad \text{"translated" topology.} \end{cases}$$

(We use the *same* notation as in (9.3), although we evidently deal with different spaces here!).

Then:

$$(12.28)\quad -\mathrm{i}\, A^* + D_t \quad \text{is an isomorphism of} \quad X^r(Q) \to H^{2(r-1)m,r}_{0,0}(Q)$$

and therefore

$$(12.29)\quad \begin{cases} \textit{for every continuous antilinear form } v \to L(v) \textit{ on } X^r(Q), \\ \textit{there exists a unique } u \in H^{-2(r-1)m,-r}(Q) \\ \textit{which is a solution of} \\ \langle u, \overline{(-\mathrm{i}\, A^* - D_t)\, v}\rangle = L(v), \quad \forall v \in X^r(Q). \end{cases}$$

12.5 Choice of L

According to Theorem 12.2, we have

$$(12.30)\quad X^r(Q) \subset H^{2rm,r}(Q).$$

Therefore, if we take

$$(12.31)\quad L(v) = \langle f, \bar{v}\rangle + \mathrm{i}\langle \vec{g}, \overline{\vec{T} v}\rangle + \langle u_0, \overline{v(0)}\rangle,$$

we see that we can choose (notation of Section 10):

$$(12.32)\quad f \in \Xi^{-2rm,-r}(Q),$$

$$(12.33)\quad \vec{g} \in \prod_{j=0}^{m-1} H^{-2rm+(2m-m_j-1/2)}\, \Xi^{-r+(2m-m_j-1/2)/2m}(\Sigma),$$

$$(12.34)\quad u_0 \in \Xi^{-(2r-1)m}(\Omega). \quad \square$$

12.6 Trace Theorem

We can (however, keeping in mind the remarks made in Section 11.6 and in Section 12.3 of Chapter 4) interpret problem (12.29) in the "usual" form:

$$(12.35)\quad \begin{cases} \mathrm{i}\, A\, u + u' = f \\ \vec{B}\, u = \vec{g} \\ u(0) = u_0, \end{cases}$$

with the following "trace theorems" (proofs analogous to those of Section 11); first of all, let:

(12.36) $D_{iA+D_t}^{-(2r-1)}(Q) = \{u \in H^{-2(r-1)m,-r}(Q), iA\,u + u' \in \Xi^{-2rm,-2r}(Q)\}$;

note that

(12.37) $\begin{cases} \text{if } u \text{ satisfies (12.29), then } u \text{ belongs to the closure} \\ \mathbf{D}_{iA+D_t}^{-(2r-1)}(Q) \text{ of } \mathscr{D}(\bar{Q}) \text{ in } D_{iA+D_t}^{-(2r-1)}(Q). \end{cases}$

Then:

(12.38) $\begin{cases} \textit{if } r \textit{ is an integer, } r \geq 1, \textit{ with } r + 1 - \left(\frac{r+1}{r}\right)\left(\frac{k+\frac{1}{2}}{2m}\right) \neq \\ \neq \textit{integer} + \frac{1}{2}, \textit{ for } 0 \leq k \leq 2rm - 1, \textit{ then the mapping} \\ u \to \vec{B} u \quad \textit{of} \quad \mathscr{D}(\bar{Q}) \to \mathscr{D}(\bar{\Sigma})^m \\ \textit{extends by continuity to a continuous linear mapping, still} \\ \textit{denoted by } u \to \vec{B} u, \textit{ of } \mathbf{D}_{iA+D_t}^{-(2r-1)}(Q) \textit{ into} \\ \prod_{j=0}^{m-1} H^{-(2rm+(2m-m_j-1/2),-(r+1))+\left(\frac{r+1}{r}\right)\left(\frac{2m-m_j-1/2}{2m}\right)}(\Sigma), \end{cases}$

and

(12.39) $\begin{cases} \textit{if } r \textit{ is an integer, } r \geq 1, \textit{ with } 2rm - \frac{r}{r+1} \neq \textit{integer} + \frac{1}{2}, \\ \textit{then the mapping } u \to u(0) \textit{ of } \mathscr{D}(\bar{Q}) \to \mathscr{D}(\bar{\Omega}) \textit{ extends by conti-} \\ \textit{nuity to a continuous linear mapping, still denoted by } u \to u(0), \\ \textit{of } \mathbf{D}_{iA+D_t}^{-(2r-1)}(Q) \quad \textit{into} \quad H^{2rm+r/(r+1)}(\Omega). \quad \Box \end{cases}$

Remark 12.3. It is possible to develop a remark entirely analogous to Section 11.6, in connection with the trace theorems for Schroedinger's equation.

13. Comments

The operators studied in Sections 1 to 11 of this chapter are *hyperbolic* if A is of order 2 and well-posed in the sense of Petrowski if A is of order $2m$, $m > 1$.

Our objective is in no way the study of the particular properties due to the hyperbolic nature of the problems. For the study of mixed (homogeneous) problems pertaining to second order hyperbolic equations, we refer the reader to Hadamard [1], Krzyzanski-Schauder [1], Schauder [1] and Ladyzenskaya [1]. For the Cauchy problem for hyperbolic systems, see Gårding [3], Leray [1], Petrowski [2], the works of Gårding and Leray, as well as (application of singular integral methods) Calderon [6], Mizohata [1, 3, 4].

For the case of constant coefficients and the use of the Fourier transform, see Schwartz [3], Gelfand-Shilov [1], Volume III, and for the use of the Laplace transform, see Garnir [1].

For mixed hyperbolic problems of order >2, for which many open questions still remain, see Agmon [5], Campbell [1], Campbell-Robinson [1], Duff [1], Thomée [1] and Peyser [1].

For first-order hyperbolic systems, see Agranovich [2, 3], De Prima [1], Friedrichs [3], Friedrichs-Lax [1], Lax-Phillips [1], Phillips [1], Phillips-Sarason [1], Sarason [1, 2], Wilcox [1, 2].

Curiously enough, the non-homogeneous boundary value problems do not seem to have undergone any systematic study for the cases considered in this chapter, even for second-order hyperbolic operators (except, of course, where only "regular" data on the boundary are considered, and where a "loss of regularity" in the result is accepted; see V. A. Il'in [1], Vaghi [1]).

It was therefore necessary to take up this question from the beginning; we start with the proof of regularity results, which (although far from optimal, see the Remarks of Section 6) seem to be new.

Once the regularity theorems have been established (Sections 2 and 7), we apply the general procedure followed in this text: we transpose and interpolate (or vice-versa).

Unfortunately, here again, "exceptional parameters" leading to serious technical difficulties appear.

We consider the variational setting; here, we can make a remark analogous to Chapter 2, Section 9 and to Chapter 4, Section 17; if we consider the problem

$$\Delta^2 u + \frac{\partial^2 u}{\partial t^2} = f, \tag{1}$$

$$\begin{cases} u(x,0) = u_0, \\ u'(x,0) = u_1, \end{cases} \tag{2}$$

$$u|_\Sigma = 0, \quad \left.\frac{\partial \Delta u}{\partial \nu}\right|_\Sigma = 0, \tag{3}$$

then we can treat it by the variational method by taking

$$H = H_0^1(\Omega), \quad V = \left\{ v \mid v \in H_0^1(\Omega), \frac{\partial}{\partial x_i} \Delta v \in L^2(\Omega) \right\},$$

$$a(t; u, v) = a(u, v) = \sum_{i=1}^n \left(\frac{\partial}{\partial x_i} \Delta u, \frac{\partial}{\partial x_i} \Delta v \right)_{L^2(\Omega)}.$$

The corresponding non-homogeneous boundary value problems have not been studied here. For other boundary value problems in connection with (1), see Prouse [1].

An analogous remark holds if we replace (1) with $\mathrm{i}\,\Delta^2 u + \frac{\partial u}{\partial t} = f$, (2) with $u(x, 0) = u_0$ and leave (3) as it stands.

In the preceding Chapters (as we have already pointed out) we could have taken regularity theorems in L^p, $p \neq 1, 2, \infty$, as a starting point. No doubt, this would have led to rather considerable additional technical difficulties and to certain open problems (which we have pointed out), but the starting point (regularity theorems) was accessible. This is no longer true for the present case; the "L^p-inequalities", which one could formally expect, are inaccurate (see Littman [1]); substitutes can been found (see Giusti-Da Prato [1], Da Prato [6]), but much work remains to be done in this direction, and we have not undertaken it.

We have purposely left aside a certain number of cases which are probably accessible, but which would have considerably complicated the presentation: transmission problems, coupled problems, etc. ...; we refer to these cases in some of the problems which follow.

For almost-periodic solutions in t, see — as was already pointed out in the Comments to Chapter 3 — the text of Amerio-Prouse [1] and its bibliography (it is likely that the almost-periodic properties of the solution extend to the various classes given in this chapter, in particular when the boundary data g are almost-periodic in t in suitable spaces $H^{-s}(\Gamma)$; but it seems that this point has not been studied to date).

For hyperbolic operators of the form $\frac{\partial^2}{\partial t^2} + A$, where A is an elliptic operator of the second order which can degenerate, see Oleinik [3, 4].

For hyperbolic operators of the form $\frac{\partial^2}{\partial t^2} + \frac{k}{t}\frac{\partial}{\partial t} + A$, see Weinstein [2] and also Diaz [1], Walter [1] and the bibliographies of these works.

14. Problems

14.1 More complete study of the traces of u's belonging to $\mathbf{D}_{A+D_t^2}^{-(2r-1)}(Q)$.

This problem is more difficult than for the parabolic case (see Section 12.3 and the Problems for Chapter 4), since, here, we do not have "optimal" estimates (see also Section 11.6 and Problem 14.5 below).

14.2 Consider the spaces A_i described in (5.2) and (5.3). It seems likely to us, but it has not been demonstrated ,that

$$[A_0, A_1]_\theta = \prod_{i=0}^{m-1} \mathscr{H}^{3m(1-\theta)-(m_j+1/2),(3m(1-\theta)-(m_j+1/2))/m}(\Sigma) \times \\ \times \mathscr{H}^{2m(1-\theta)-\theta m/2}(\Omega) \times \mathscr{H}^{(3m/2)(1-2\theta)}(\Omega)$$

if

$$\frac{3m(1-\theta)-(m_j+\frac{1}{2})}{m}<\frac{1}{2}, \tag{14.1}$$

and with the $\mathcal{R}.\mathcal{C}.$ *being well-defined* if (14.1) is not satisfied.

14.3 Characterize (in a different way than by the definition!) the space

$$[H^{0,1}(Q), (H^{2m,2}(Q))']_\theta$$

(which appears in (5.5), for instance).

14.4 Because we wanted to remain within a *Hilbert* theory, our regularity results are never *optimal*. For example, in Theorem 2.1, we may conclude that

$$u \in L^\infty(0, T; H^{2m}(\Omega)),$$
$$u'' \in L^\infty(0, T; H^0(\Omega)),$$

(instead of $L^2(0, T; H^{2m}(\Omega))$ and $L^2(0, T; H^0(\Omega))$). Therefore, by transposition more precise results can be obtained by using a non-Hilbert theory.

In this way, *problems of interpolation between subspaces of Sobolev spaces with different exponents* arise.

The following is an example of such problems: consider an open set

$$\Omega \times \mathcal{O}, \quad \Omega \in \mathbf{R}^n_x, \quad \mathcal{O} \in \mathbf{R}^m_y,$$

and first let

$$W^\mu_p(\Omega; W^\nu_q(\mathcal{O})) = \{v \mid D^\alpha_x v \in L^p(\Omega; W^\nu_q(\mathcal{O})), |\alpha| \leqq \mu\},$$

where

$$W^\nu_q(\mathcal{O}) = \{w \mid D^\beta_y w \in L^q(\mathcal{O}), |\beta| \leqq \nu\};$$

next, let

$V = \{v \mid v \in W^\mu_p(\Omega; W^\nu_q(\mathcal{O})), B_j v = 0$ on all (or a subset) of the boundary of $\Omega \times \mathcal{O}$, the B_j's being suitable boundary operators$\}$.

Can the spaces $(V, L^p(\Omega; L^q(\mathcal{O})))_{\theta,r}$ (in the notation of Peetre [8]) be characterized?

14.5 Is $\mathcal{D}(\bar{Q})$ dense in $D^{-(2r-1)}_{A+D^2_t}(Q)$ (Theorem 11.1)? In this connection, see Baiocchi [8] and Hörmander [4, 5].

14.6 It might be useful to study the "non-integer r" cases and, in particular, the exceptional values of r; a study which we have not attempted.

14.7 Non-homogeneous, *general hyperbolic* problems; the theory of mixed problems (in the sense of Hadamard) for these cases is still wide open; see the Comments.

14.8 Consider the operators

$$A\left(x,t,\frac{\partial}{\partial x}\right) + A_1\left(x,t,\frac{\partial}{\partial x}\right)\frac{\partial}{\partial t} + A_2\left(x,t,\frac{\partial}{\partial x}\right)\frac{\partial^2}{\partial t^2} \tag{14.2}$$

under various hypotheses on A_i (see Lions [13]; Baiocchi [6, 7]; see also Chapter 3).

It should be possible, without too much difficulty, to extend the results of this chapter to the operators (14.2). See also the study of operators of type (14.2) in Albertoni/Cercignani [1] (see also Scarpini [1]).

14.9 Similarly, it should be possible to extend the results of this chapter to coupled problems: on an open set $\Omega_1 \times \Omega_2$, we have an operator of a given type – for example parabolic – on $\Omega_1 \times]0, T[$ and of *another type*, for example hyperbolic, etc., on $\Omega_2 \times]0, T[$, with *transmission conditions* on $\Gamma_1 \times]0, T[$, where Γ_1 is a *common part* of the boundaries of Ω_1 and Ω_2.

For homogeneous problems of this type, see Gelfand [1], Hersch [1], Lions [18], Lions-Raviart [1].

14.10 We have always assumed that the operators A (or A_i in (14.2)) are "regular". Study the non-homogeneous boundary value problems for the operators $A + \frac{\partial^2}{\partial t^2}$, where A *has singularities*. For example: $A = -\Delta - \frac{k}{x_n}\frac{\partial}{\partial x_n}$, where Ω is contained in $\{x_n > 0\}$, with part of the boundary on the (singular) hyperplane $x_n = 0$. Or, where A can *degenerate*; see Baouendi [1] and studies by O. A. Oleinik which are still to be published.

14.11 Regularity theorems and non-homogeneous problems for the evolution operators corresponding to the semi-groups of order n of Da Prato [1, 2].

14.12 Non-homogeneous problems for, for example,

$$A + \frac{\partial^2}{\partial t^2}$$

on $\Omega \times]0, T[$, where Ω has various parts of the boundary Γ_i of *different dimensions* (analogue to Problem 17.7, Chapter 4).

14.13 Analogue to Problem 17.13, Chapter 4, with the "elliptic regularization" connected with $A + \frac{\partial^2}{\partial t^2}$ (see Strauss [1]).

14.14 Analogue to the preceding problem with the "parabolic regularization" introduced in Section 12.2, and the one of Chapter 3, Section 8. (In this connection, note that, having applied the "parabolic regularization"

$$\mathrm{i}\, A + \frac{\partial}{\partial t} \Rightarrow (\varepsilon + \mathrm{i})\, A + \frac{\partial}{\partial t},$$

we can apply the "elliptic regularization to the parabolically regularized operator", to obtain

$$\left((\varepsilon + \mathrm{i})\, A + \frac{\partial}{\partial t}\right) - \varepsilon_1 \frac{\partial^2}{\partial t^2}, \qquad \varepsilon_1 > 0,$$

or, if A is of order $2m$:

$$(\varepsilon + \mathrm{i})\, A + \frac{\partial}{\partial t} + \varepsilon_1 (-1)^m \cdot \frac{\partial^{2m}}{\partial t^{2m}}. \tag{14.3}$$

In this way, we are led back to solving non-homogeneous problems for the elliptic operators (14.3), followed by results of the type discussed in Section 12, as $\varepsilon, \varepsilon_1 \to 0$).

14.15 Problems of the preceding type in *non-cylindrical* domains; see Rogak [1] and Lions [31].

14.16 Which non-homogeneous problems for the operators considered in this chapter can be solved in L^p, $p \neq 2$? There are essential differences between the cases $p = 2$ and $p \neq 2$ (see Littman [1]), but the results of Da Prato-Giusti [1] and Da Prato [6] should serve as a "starting point" for a study of these problems.

14.17 Study of coupled problems (f and g being taken from very general classes) of the type

$$-\Delta_x u = f(x, t) \quad \text{in} \quad Q,$$

$$\frac{\partial u}{\partial \nu} + \frac{\partial^2 u}{\partial t^2} = g \quad \text{on} \quad \Sigma,$$

$$u(x, 0) = u_0 \quad \text{for} \quad x \in \Omega,$$

$$\frac{\partial u}{\partial t}(x, 0) = u_1 \quad \text{for} \quad x \in \Gamma,$$

(see Problem 17.12 of Chapter 4); see Friedman-Shimbrot [1], Lions [13].

14.18 The problem of *oblique derivatives* for the *wave operator* seems to be unsolved. Is it well-posed?

14.19 The problems of this chapter can be studied in weighted H^m-spaces (for the elliptic case, see, in particular, Geymonat-Grisvard [2]).

More generally, can these problems be studied in Morrey type spaces? As we have already pointed out, the "natural" inequalities are inaccurate in L^p, but do there exist (non-hilbertisable) Banach spaces where the "natural" inequalities hold?

14.20 The questions pertaining to approximation by finite difference methods for the *non-homogeneous* problems studied in this chapter lead to numerous, still unsolved problems.

Chapter 6

Applications to Optimal Control Problems

Note. This chapter is of a somewhat different nature than the others: its sole objective is to serve as an *introduction* to the (enormous) subject of control problems for systems governed by partial differential equations, with the controls *appearing* (mostly) *in the boundary conditions*. Sections 1 through 9 rely in an essential way on Chapter 4, and the remaining sections on Chapter 5[1].

We call attention to the notation, as it is *different* than for Chapters 4 and 5; we have, whenever possible, followed the usual notation of control theory.

1. Statement of the Problems for the Linear Parabolic Case

1.1 Notation

In this chapter, we adopt, as much as possible, the *usual notation of control theory*, which is different than the notation for the theory of boundary value problems (and thus from the one used up to this point).

For the *geometric* concepts the notation remains unchanged: Ω, Γ, Σ, etc.

Similarly, we still have: $x \in \Omega$, $t =$ time, etc.

The state of the system is represented by $y(x, t)$, a function taking its values in $\mathbf{R}$ (or in $\mathbf{R}^N$ for "differential systems").

Every function under consideration takes its values in $\mathbf{R}$; *thus the scalar products are real, and denoted by* $\langle\ ,\ \rangle$ *or* $(\ ,\ )$, *without distinction.* It is assumed that the state of the system y is provided by the solution of the *non-homogeneous boundary value problem*:

(1.1) $$A\,y + y' = E_1\,u_1 \quad \text{in} \quad Q = \Omega \times]0, T[,$$

(1.2) $$\vec{B}\,y = E_2\,u_2,$$

(1.3) $$y(x, 0) = E_3\,u_3,$$

[1] Nevertheless, Section 1 may be read independently from the rest of the text. The fractional Sobolev spaces intervene (as trace spaces) in the following sections.

where

(i) $A + \frac{\partial}{\partial t}$ is a *parabolic operator satisfying the hypotheses* of Section 1, Chapter 4, for the operators $\{A, B_j\}$, with $\vec{B} = \{B_j\}$;
(ii) u_1, u_2, u_3 *denote the controls.* □

More precisely, we assume that

$$u_i \in \mathscr{H}_i = \text{Hilbert space on } \mathbf{R}, \quad i = 1, 2, 3 \tag{1.4}$$

(with scalar product $(\ ,\ )_{\mathscr{H}_i}$) and that

$$\begin{cases} E_1 \in \mathscr{L}(\mathscr{H}_1; L^2(Q)), \\ E_2 \in \mathscr{L}(\mathscr{H}_2; L^2(\Sigma)^m), \\ E_3 \in \mathscr{L}(\mathscr{H}_3; L^2(\Omega)). \end{cases} \tag{1.5}$$

Then, for $u = \{u_1, u_2, u_3\}$ given in $\prod_{i=1}^{3} \mathscr{H}_i$, problem (1.1), (1.2), (1.3) admits a unique solution (see Chapter 4; we shall come back to this point):

$$y(x, t) = y(x, t; u), \quad u = \{u_1, u_2, u_3\}. \tag{1.6}$$

Remark 1.1. Of course, we could consider an *infinity* of other situations: the operators E_i mapping $\mathscr{H}_i$ into spaces different from $L^2(Q)$, etc. (for example, $H^{2rm,r}(Q)$, $\Xi^{-2rm,-r}(Q)$ instead of $L^2(Q)$, etc.); note that:

(i) setting (1.5) seems to be the most important one for applications;
(ii) the method developed below is *general.* □

Remark 1.2. We could also consider

$$A y + y' = E_1 u_1 + f, \tag{1.7}$$

$$\vec{B} y = E_2 u_2 + \vec{g}, \tag{1.8}$$

$$y(x, 0) = E_3 u_3 + y_0(x), \tag{1.9}$$

with *given* f, $\vec{g}$, y_0. Everything that follows can be adapted to this situation, without difficulty. □

An important case for the applications is when f, $\vec{g}$, y_0 are *random functions.* We shall not develop this point of view here; see Bensoussan [1]. □

Remark 1.3. We insist on the fact that T is *fixed*; we do not consider optimal time problems here. □

1.2 Optimization Problems

1.2.1 The constraints on the control

The controls u_i do not always describe the entire space $\mathscr{H}_i$, but

(1.10) $$u = \{u_1, u_2, u_3\} \text{ describes } \mathscr{U} \subset \vec{\mathscr{H}} = \prod_{i=1}^{3} \mathscr{H}_i.$$

We shall assume that

(1.11) $\mathscr{U}$ is a *closed, convex subset of* $\mathscr{H}$.

1.2.2 The cost function $J(u)$

Let us begin with a "vague" statement of the problem: as u describes $\mathscr{U}$, $y(x, t; u)$ describes a set of functions; we seek u (or the u's) (if it exists) such that $y(x, t; u)$ is "as close as possible" to a *desired state* $y_d(x, t)$.

It is not necessary, but convenient (and useful for the applications) to consider $y(x, t; u)$:

1) in Q;

2) on Σ, *via* $\vec{S}\, y$ (see Chapter 4, Section 1; it is natural to work with $\vec{S}\, y$, since $\vec{B}\, y$ is *given* and since $\{\vec{B}, \vec{S}\}$ is a Dirichlet system);

3) on Ω_T, that is $y(x, T; u)$.

We thus seek (and this is still not a precise statement) u in $\mathscr{U}$ such that $y(x, t; u)$ is "as close a possible" to y_d^1 in Q, $\vec{S}\, y$ is "as close as possible" to y_d^2 on Σ, $y(x, T; u)$ is "as close as possible" to y_d^3 in Ω_T. □

Now, let us make the *sense of this approximation* precise.

Let K_i, $i = 1, 2, 3$, be three Hilbert spaces of functions or distributions on Q, Σ, Ω, respectively. For the sake of precision, we shall sometimes write

$$K_1(Q), \quad K_2(\Sigma), \quad K_3(\Omega).$$

We assume:

(1.12) $$y_d^i \text{ given in } K_i, \quad i = 1, 2, 3.$$

The essential hypothesis is:

(1.13) $$\begin{cases} \textit{For } y(x, t; u), \textit{ solution of } (1.1), (1.2), (1.3), \textit{ the spaces } K_i \textit{ must} \\ \textit{be chosen such that } y \in K_1(Q),\ \vec{S}\, y \in K_2(\Sigma),\ y(x, T) \in K_3(\Omega). \end{cases}$$

(This shall be made precise in Section 2.) □

It is then natural to seek a solution "as close as possible" to y_d^i *in the sense of* K_i. From which we obtain the following precise definitions.

Definition 1.1. *The cost function is the functional:*

$$J(u) = \beta_1 \|y(x,t;u) - y_d^1\|_{K_1}^2 + \beta_2 \|\tilde{S}\, y - y_d^2\|_{K_2}^2 + \beta_3 \|y(x,T;u) - y_d^3\|_{K_3}^2, \tag{1.14}$$

where

$$\beta_i \geqq 0, \quad \beta_1 + \beta_2 + \beta_3 > 0.$$

The control (or optimization) problem is to seek the u^0 (or u^0's) in $\mathscr{U}$ (if it exists) such that

$$J(u^0) \leqq J(u), \quad \forall u \in \mathscr{U}. \tag{1.15}$$

Definition 1.2. Every control u^0 satisfying (1.15) is said to be an "*optimal control*".

Remark 1.4. Up to a certain point, we are free to choose the spaces K_i (see below); the essential point is that these spaces *cannot* be chosen "arbitrarily small". For example, we may think that we can seek an "optimal control" for

$$\beta_3 \int_\Omega |y(x,T;u) - y_d^3|^2\, dx,$$

with y_d^3 given in $L^2(\Omega)$. But under hypotheses (1.5), *this problem is meaningless* (except if $\inf. m_j = m$); we *must* take (except if $\inf. m_j = m$) $K_3(\Omega)$ larger than $L^2(\Omega)$. □

Remark 1.5. We restrict our study to quadratic functionals of type (1.14). Of course, one could *state* the same kind of problem, replacing the spaces $L^2(Q)$ with $L^p(Q)$, $1 \leqq p \leqq \infty$, $L^2(\Sigma)$ with $L^p(\Sigma)$, etc. and using other, non-quadratic functionals. □

Remark 1.6. The cost function is generally written in terms of *observations*, which are themselves *linear functions of the state* y. Up to certain technical details, the following methods remain valid for this case. □

2. Choice of the Norms in the Cost Function

2.1 Reminder. Condition on $K_1(Q)$

According to Theorem 15.2, Chapter 4, we have:

$$y \in H^{\mu,\mu/2m}(Q), \tag{2.1}$$

where

$$\mu = \tfrac{1}{2} + \min_j m_j, \tag{2.2}$$

except if $\mu = m + \frac{1}{2}$, i.e. if $\min m_j = m$. (Indeed, for this case, Theorem 15.2, Chapter 4, assumes that $y(x, 0)$ is given in $H^{1/2}(\Omega)$ and not only in $L^2(\Omega)$. But (2.1) is still valid if $E_3 = 0$ (*or if* $E_3 \in \in \mathscr{L}(\mathscr{H}_3; H^{1/2}(\Omega))$).)

In the sequel, we shall assume that $\mu < m + \frac{1}{2}$ *and that* $E_3 = 0$ if $\mu = m + \frac{1}{2}$. (The case "$\mu = m + \frac{1}{2}$, $E_3 \neq 0$" is studied in Section 5.)

Under these conditions $\left(\text{and assuming that } u \text{ describes the } \textit{entire} \text{ space } \vec{\mathscr{H}} = \prod_{i=1}^{3} \mathscr{H}_i\right)$, we have:

Proposition 2.1. $K_1(Q)$ (see 1.2.2) *must be chosen to satisfy*

$$K_1(Q) \supset H^{\mu,\mu/2m}(Q). \quad \square \tag{2.3}$$

Furthermore, we shall assume that

$$H^{\mu,\mu/2m}(Q) \quad \text{is dense in} \quad K_1(Q). \tag{2.4}$$

2.2 Space Described by $\vec{S}y$. Condition on $K_2(\Sigma)$

We know (Chapter 4, Section 2) that, if μ_j denotes the order of S_j,

$$\begin{cases} y \to \vec{S}y \quad \text{is a continuous linear mapping of} \\ H^{2m,1}(Q) \to \prod_{j=0}^{m-1} H^{2m-\mu_j-1/2,(2m-\mu_j-1/2)/2m}(\Sigma) \end{cases} \tag{2.5}$$

and that (Chapter 4, Section 10; Remark 10.1)

$$\begin{cases} y \to \vec{S}y \quad \text{is a continuous linear mapping of} \\ D_P^0(Q) \to \prod_{j=0}^{m-1} H^{-\mu_j-1/2,(-\mu_j-1/2)/2m}(\Sigma). \end{cases} \tag{2.6}$$

But $y \in H^{\mu,\mu/2m}(Q)$ *and* $Py = Ay + y' \in L^2(Q)$, so that, by interpolation of (2.5), (2.6), we obtain:

$$\begin{cases} \text{if } y \in H^{\mu,\mu/2m}(Q) \quad \text{and} \quad Py \in L^2(Q), \quad \text{then} \\ \vec{S}y \in \prod_{j=0}^{m-1} H^{\mu-(2m-\mu_j-1)-1/2,(\mu-(2m-\mu_j-1)-1/2)/2m}(\Sigma). \end{cases} \tag{2.7}$$

Therefore:

Proposition 2.2. $K_2(\Sigma)$ *must be chosen to satisfy*

$$K_2(\Sigma) \supset \prod_{j=0}^{m-1} H^{\mu-\mu_j-1/2,(\mu-\mu_j-1/2)/2m}(\Sigma). \tag{2.8}$$

Furthermore, as in (2.4), we shall assume that

(2.9) *in* (2.8), *the second space is dense in the first.*

2.3 Space Described by $y(x, T; u)$. Condition on $K_3(\Omega)$

We have (2.1), therefore

$$y \in L^2(0, T; H^\mu(\Omega)).$$

According to Theorem 12.7, Chapter 1, we therefore have (since $\mu = \text{integer} + \frac{1}{2}$)

$$A\, y \in L^2(0, T; (H_{00}^{2m-\mu}(\Omega))')$$

and therefore

$$y' = E_1 u_1 - A\, y \in L^2(0, T; (H_{00}^{2m-\mu}(\Omega))').$$

Then (Chapter 1, Sections 3 and 9)

$$y(., T; u) \in [H^\mu(\Omega), (H_{00}^{2m-\mu}(\Omega))']_{1/2} = X;$$

by the duality theorem, Chapter 1, Section 6.2:

$$X' = [H_{00}^{2m-\mu}(\Omega), (H^\mu(\Omega))']_{1/2}$$

and according to Remark 12.6, Chapter 1, we have

$$X' = \begin{cases} H_{00}^{m-\mu}(\Omega) & \text{if } \mu \leqq m - \frac{1}{2}, \\ H^{m-\mu}(\Omega) & \text{if } \mu = m + \frac{1}{2}. \end{cases}$$

Therefore

$$\text{(2.10)} \qquad y(., T; u) \begin{cases} \in (H_{00}^{m-\mu}(\Omega))' & \text{if } \mu \leqq m - \frac{1}{2} \\ \in H^{1/2}(\Omega) & \text{if } \mu = m + \frac{1}{2}. \end{cases}$$

Therefore

Proposition 2.3. *$K_3(\Omega)$ must be chosen to satisfy*

$$\text{(2.11)} \qquad \begin{cases} K_3(\Omega) \supset (H_{00}^{m-\mu}(\Omega))' & \text{if } \mu \leqq m - \frac{1}{2}, \\ K_3(\Omega) \supset H^{1/2}(\Omega) & \text{if } \mu = m + \frac{1}{2}. \end{cases}$$

And, as in (2.4) and (2.9), we add

(2.12) $(H_{00}^{m-\mu}(\Omega))'$ (or $H^{1/2}(\Omega)$) *is dense in* $K_3(\Omega)$. □

Remark 2.1. Of course, conditions (2.3), (2.8) and (2.11) *do not uniquely define* the K_i's! *There exists an infinity of possible choices.* □

Remark 2.2. If

$$K_2 = \prod_{j=0}^{m-1} K_{2,j}, \quad \text{with} \quad \|z\|_{K_2}^2 = \sum_{j=0}^{m-1} \|z_j\|_{K_{2,j}}^2,$$

than we can consider

$$\sum_{j=0}^{m-1} \beta_{2j} \|(\tilde{S}\, y)_j - y_{d,j}^2\|_{K_{2,j}}^2$$

instead of $\beta_2 \|\tilde{S}\, y - y_d^2\|_{K_2}^2$, in (1.14). □

3. Optimality Condition for Quadratic Cost Functions

In this section, we recall a well-known result.

3.1 Notation

Let $\pi(u, v)$ be a continuous bilinear form on a real Hilbert space $\mathscr{H}$ and let $v \to M(v)$ be a continuous linear form on $\mathscr{H}$.

Let $\mathscr{U}$ be a *closed convex* subset of $\mathscr{H}$.

Set

$$J(u) = \pi(u, u) - 2M(u). \tag{3.1}$$

We seek the u^0 (or u^0's) *in* $\mathscr{U}$ (if it exists) such that

$$J(u^0) \leqq J(u), \quad \forall u \in \mathscr{U}. \tag{3.2}$$

3.2 Optimality Condition

Proposition 3.1. *Assume that* $\pi(u, v)$ *satisfies:*

$$\pi(u, v) = \pi(v, u), \quad \forall u, v \in \mathscr{H}, \tag{3.3}$$

and

$$\pi(u, u) \geqq 0, \quad \forall u \in \mathscr{H}. \tag{3.4}$$

Then, a necessary and sufficient condition for (3.2) *to be satisfied is that*

$$\pi(u^0, u - u^0) \geqq M(u - u^0), \quad \forall u \in \mathscr{U}. \tag{3.5}$$

Proof. 1) (3.2) $\Rightarrow$ (3.5).

Let u be arbitrary in $\mathscr{U}$; then $(1 - \theta) u^0 + \theta u \in \mathscr{U}$ for $0 < \theta < 1$ and therefore (3.2) implies

$$J(u^0) \leqq J((1 - \theta) u^0 + \theta u),$$

that is, expanding (and using the symmetry (3.3)),

$$0 \leqq 2\theta [\pi(u^0, u - u^0) - M(u - u^0)] + \theta^2 \pi(u - u^0, u - u^0).$$

Dividing by θ and letting $\theta \to 0$, we obtain (3.5).

2) (3.5) $\Rightarrow$ (3.2).

For u (and u^0) arbitrary in $\mathscr{U}$, we have:

$$J(u) - J(u^0) = 2[\pi(u^0, u - u^0) - M(u - u^0)] + \pi(u - u^0, u - u^0), \tag{3.6}$$

from which the result follows, since $\pi(u - u^0, u - u^0) \geqq 0$, by (3.4). □

4. Optimality Condition and Green's Formula

4.1 Optimality Condition. Application of Section 3.2

We apply the remarks of Section 3 to the problem stated in Sections 1 and 2. Expanding (1.14), we may write

$$J(u) = \pi(u, u) - 2M(u) + \text{constant}, \tag{4.1}$$

where

$$\begin{cases} \pi(u, u) = \beta_1(y(x, t; u), y(x, t; u))_{K_1} + \beta_2(\vec{S}\, y(x, t; u), \\ \vec{S}\, y(x, t; u))_{K_2} + \beta_3(y(x, T; u), y(x, T; u))_{K_3}, \end{cases} \tag{4.2}$$

$$\begin{cases} M(u) = \beta_1(y(x, t; u), y_d^1)_{K_1} + \beta_2(\vec{S}\, y(x, t; u), y_d^2)_{K_2} + \\ \qquad + \beta_3(y(x, T; u), y_d^3)_{K_3} \end{cases} \tag{4.3}$$

and where the constant equals

$$\sum_{i=1}^{3} \beta_i \, \| y_d^i \|_{K_i}^2 .$$

Since

$$u = \{u_1, u_2, u_3\} \to \{y(x, t; u), \vec{S}\, y, y(x, T; u)\}$$

is a *continuous linear* mapping of

$$\vec{\mathscr{H}} = \prod_{i=1}^{3} \mathscr{H}_i \to K_1 \times K_2 \times K_3$$

(see Propositions 2.1, 2.2. and 2.3) we may apply Proposition 3.1 to $J(u)$. We thus obtain:

Proposition 4.1. *A necessary and sufficient condition for u^0 to be an optimal control is that* (we have written $y(u)$ for $y(x, t; u)$):

$$\begin{cases} \beta_1(y(u^0) - y_d^1, y(u) - y(u^0))_{K_1} + \\ + \beta_2(\vec{S}\, y(u^0) - y_d^2, \vec{S}\, y(u) - \vec{S}\, y(u^0))_{K_2} + \\ + \beta_3(y(x, T; u^0) - y_d^3, y(x, T; u) - y(x, T; u^0))_{K_3} \geqq 0. \\ \forall u \in \mathscr{U}. \quad \square \end{cases} \tag{4.4}$$

We shall now transform this condition by using Green's formula. To this end, we need some preparation.

4.2 The Isomorphisms Λ_i

We distinguish two cases for the choice of K_i:

First case: $\mu \leqq m - \frac{1}{2}$. Then, we use (2.3), (2.8) and (2.11), to obtain:

$$(4.5)_1 \qquad K_1'(Q) \subset (H^{\mu,\mu/2m}(Q))',$$

$$(4.6)_1 \qquad K_2'(\Sigma) \subset \prod_{j=0}^{m-1} (H^{\mu-\mu_j-1/2,(\mu-\mu_j-1/2)/2m}(\Sigma))',$$

$$(4.7)_1 \qquad K_3'(\Omega) \subset H_{00}^{m-\mu}(\Omega).$$

Second case: $\mu = m + \frac{1}{2}$. Then, so as to avoid having to use the boundary value problems for exceptional values of the parameter, we choose $\varepsilon > 0$, fixed arbitrarily small, and replace (2.3), (2.8) and (2.11) with

$$K_1(Q) \supset H^{\mu-\varepsilon,(\mu-\varepsilon)/2m}(Q),$$

$$K_2(\Sigma) \supset \prod_{j=0}^{m-1} H^{\mu-\varepsilon-\mu_j-1/2,(\mu-\varepsilon-\mu_j-1/2)/2m}(\Sigma),$$

$$K_3(\Omega) \supset H^{1/2-\varepsilon}(\Omega).$$

Then

$$(4.5)_2 \qquad K_1'(Q) \subset (H^{\mu-\varepsilon,(\mu-\varepsilon)/2m}(Q))',$$

$$(4.6)_2 \qquad K_2'(\Sigma) \subset \prod_{j=0}^{m-1} (H^{\mu-\varepsilon-\mu_j-1/2,(\mu-\varepsilon-\mu_j-1/2)/2m}(\Sigma))',$$

$$(4.7)_2 \qquad K_3'(\Omega) \subset (H^{1/2-\varepsilon}(\Omega))' = H^{-1/2+\varepsilon}(\Omega). \quad \square$$

Definition 4.1. *We denote by Λ_i the canonical isomorphism of K_i onto K_i'.*

4.3 The "Adjoint" Problem

The notation for the operators A^*, $\vec{C}$, $\vec{T}$ is the same as in Section 1 of Chapter 4.

For $u \in \vec{\mathscr{H}}$, we denote by $p(x, t; u) = p(u)$ the solution of the following ("adjoint") problem:

$$(4.8) \qquad \left(A^* - \frac{\partial}{\partial t}\right) p(u) = \beta_1 \Lambda_1 (y(u) - y_d^1),$$

$$(4.9) \qquad \vec{C} p(u) = \beta_2 \Lambda_2 (\vec{S} y(u) - y_d^2),$$

$$(4.10) \qquad p(x, T; u) = \beta_3 \Lambda_3 (y(x, T; u) - y_d^3).$$

Proposition 4.2. *Problem* (4.8), (4.9), (4.10) *admits a unique solution, which satisfies:*

$$(4.11)_1 \qquad p(u) \in H^{2m-\mu, 1-\mu/2m}(Q), \quad \text{if} \quad \mu \leqq m - \tfrac{1}{2},$$

$$(4.11)_2 \qquad p(u) \in H^{2m-\mu+\varepsilon, 1-(\mu+\varepsilon)/2m}(Q), \quad \text{if} \quad \mu = m + \tfrac{1}{2}.$$

Proof[1]. We use Chapter 4, Section 15.1, (iii), with $2\theta m = \mu$. If $y(u) - y_d^1 \in K_1(Q)$, then $\Lambda_1(y(u) - y_d^1) \in K_1'(Q)$ and (we have done what was necessary for that) $\Lambda_1(y(u) - y_d^1)$ therefore belongs to $\mathcal{H}^{-\mu,-\mu/2m}(Q)$, with the notation of Chapter 4, Sections 14 and 15.

Next

$$\Lambda_2(\tilde{S}y(u) - y_d^2) \in K_2'(\Sigma) \subset \prod_{j=0}^{m-1} \mathcal{H}^{-\mu+\mu_j+1/2,(-\mu+\mu_j+1/2)2m}(\Sigma)$$

and

$$\Lambda_3(y(x,T;u) - y_d^3) \in K_3'(\Omega).$$

Thus Chapter 4, Section 15,1. (iii) applies and we obtain (4.11). □

4.4 New Form of the Optimality Condition

Theorem 4.1. *Assume that* $\min m_j < m$ *or, if* $\min m_j = m$, *that* $E_3 \equiv 0$. *A necessary and sufficient condition for* u^0 *to be an optimal control is that:*

$$(4.12) \qquad \begin{cases} \langle p(u^0), E_1(u_1 - u_1^0)\rangle + \langle \vec{T} p(u^0), E_2(u_2 - u_2^0)\rangle + \\ \qquad + \langle p(x,0;u^0), E_3(u_3 - u_3^0)\rangle \geqq 0 \\ \qquad \forall u = \{u_1, u_2, u_3\} \in \mathcal{U}, \end{cases}$$

where $p(u^0)$ *is defined by* (4.8), (4.9), (4.10) (*with* $u = u^0$), *and where the brackets denote the scalar products in* $L^2(Q)$, $(L^2(\Sigma))^m$ *and* $L^2(\Omega)$, *respectively, since*[2]

$$p(u^0) \in H^{2m-\mu,1-\mu/2m}(Q) \subset L^2(Q),$$

$$\vec{T} p(u^0) \in \prod_{j=0}^{m-1} H^{m_j+1/2-\mu,(m_j+1/2-\mu)/2m}(\Sigma) \subset (L^2(\Sigma))^m$$

and $p(x,0;u^0) \in H^{m-\mu}(\Omega) \subset L^2(\Omega)$, *except if* $\min m_j = m$, *in which case the last scalar product in* (4.12) *drops out* (by hypothesis!).

Proof[3]. Set $\Phi = y(u) - y(u^0)$ and take the scalar product of (4.8) (with $u = u^0$) with Φ (in the duality between $K_1'(Q) \subset (H^{\mu,\mu/2m}(Q))'$

---

[1] If $\mu = m + \frac{1}{2}$, replace μ with $\mu - \varepsilon$ and $H_{00}^{m-\mu}(\Omega)$ with $H^{-1/2+\varepsilon}(\Omega)$ in the proof [see $(4.7)_2$].

[2] Here and in the sequel, replace μ with $\mu - \varepsilon$, if $\mu = m + \frac{1}{2}$.

[3] See the footnote to the proof of Proposition 4.2.

and $H^{\mu,\mu/2m}(Q)$). We obtain

$$\left\langle\left(A^* - \frac{\partial}{\partial t}\right)p(u^0), \Phi\right\rangle = \beta_1(y(u^0) - y_d^1, \Phi)_{K_1} \tag{4.13}$$

(since Λ_1 is the canonical isomorphism of K_1 onto K_1').

But, according to Green's formula, valid here by extension by continuity (see Chapter 4,) the first term in (4.13) is

$$\langle \vec{T}\, p(u^0), \vec{B}\, \Phi\rangle - \langle \vec{C}\, p(u^0), \vec{S}\, \Phi\rangle - \langle p(T; u^0), \Phi(T)\rangle + \\ + \langle p(0; u^0), \Phi(0)\rangle + \langle p(u^0), (A + \partial/\partial t)\, \Phi\rangle,$$

where the first bracket denotes the duality between

$$\prod_{j=0}^{m-1} H^{m_j+1/2-\mu,(m_j+1/2-\mu)/2m}(\Sigma) \quad \text{and} \quad (L^2(\Sigma))^m,$$

the second between K_2' and K_2, the third between K_3' and K_3 for example, the fourth between $H^{m-\mu}(\Omega) \subset L^2(\Omega)$ and $L^2(\Omega)$ (this bracket vanishing for $\mu = m + \frac{1}{2}$) and the last between $L^2(Q)$ and itself.

Using (4.9)–(4.10), it follows that:

$$\begin{cases} \beta_1(y(u^0) - y_d^1, \Phi)_{K_1} + \beta_2(\vec{S}\, y(u^0) - y_d^2, \vec{S}\, \Phi)_{K_2} + \\ + \beta_3(y(T; u^0) - y_d^3, \Phi(T))_{K_3} = \left\langle p(u^0), \left(A + \dfrac{\partial}{\partial t}\right)\Phi\right\rangle + \\ + \langle \vec{T} p(u^0), \vec{B}\Phi\rangle + \langle p(0; u^0), \Phi(0)\rangle. \end{cases} \tag{4.14}$$

Replacing $y(u) - y(u^0)$ with its value, we see that the first term in (4.4) equals

$$\left\langle p(u^0), \left(A + \frac{\partial}{\partial t}\right)\Phi\right\rangle + \langle \vec{T} p(u^0), \vec{B}\Phi\rangle + \langle p(0; u^0), \Phi(0)\rangle.$$

But

$$\left(A + \frac{\partial}{\partial t}\right)\Phi = E_1(u_1 - u_1^0) \quad (\text{if } u^0 = (u_1^0, u_2^0, u_3^0)),$$

$$\vec{B}\,\Phi = E_2(u_2 - u_2^0),$$

$$\Phi(0) = E_3(u_3 - u_3^0),$$

and the theorem follows. □

In the following sections (6 and 7), we shall give some applications of this theorem. Now, we shall briefly investigate the particular case $\mu = m + \frac{1}{2}$, $E_3 \neq 0$.

5. The Particular Case $\mu = m + \frac{1}{2}$, $E_3 \neq 0$

5.1 Properties of y

We no longer apply Theorem 15.2, Chapter 4, but instead Section 15.1 (iii) of Chapter 4, with $s = -\frac{1}{2}$; then we must have

$$E_1 u_1 \text{ (which replaces } f) \in \mathscr{H}^{-m, -1/2}(Q),$$

which is evidently satisfied, and we must also have $E_2 u_2$ (which replaces $\vec{g} = \{g_j\}$)

$$\in \prod_{j=0}^{m-1} \mathscr{H}^{m-m_j-1/2, (m-m_j-1/2)2m}(\Sigma),$$

which is satisfied since $m_j \geqq m$. In this manner, we obtain:

Proposition 5.1. *If* $\mu = m + \frac{1}{2}$, *then*

(5.1) $$y \in H^{m, 1/2}(Q). \quad \square$$

Remark 5.1. This is indeed different than (2.1). □

5.2 Choice of $K_1(Q)$

Following (5.1), we take (compare with Section 2):

(5.2) $K_1(Q) \supset H^{m, 1/2}(Q)$, the second space being dense in the first.

5.3 Choice of $K_2(\Sigma)$ and $K_3(\Omega)$

The remarks of Sections 2.2 and 2.3 hold if we replace μ with $m + \frac{1}{2}$:

(5.3) $$\begin{cases} K_2(\Sigma) \supset \prod_{j=0}^{m-1} H^{m+\mu_j, (m+\mu_j)/2m}(\Sigma), \\ \text{the second space being dense in the first.} \end{cases}$$

According to Proposition 2.3, we must take

(5.4) $$K_3(\Omega) \supset H^{1/2}(\Omega),$$

the second space being dense in the first.

In this case, a simple choice is

(5.5) $$K_3(\Omega) = L^2(\Omega). \quad \square$$

5.4 Adjoint Problem and Optimality Condition

We obtain results analogous to those of Section 4, with the new choices (5.2), (5.3) and (5.4).

6. Consequences of the Optimality Condition (I)

6.1 Generalities

Let us consider the operator E_i^*, adjoint of E_i, $i = 1, 2, 3$; with the appropriate identifications, we have:

$$(6.1)\qquad \begin{cases} E_1^* \in \mathcal{L}(L^2(Q); \mathcal{H}_1), \\ E_2^* \in \mathcal{L}((L^2(\Sigma))^m; \mathcal{H}_2), \\ E_3^* \in \mathcal{L}(L^2(\Omega); \mathcal{H}_3). \end{cases}$$

Then, if we introduce

$$(6.2)\qquad \zeta(u^0) = \{E_1^* p(u^0), E_2^* \vec{T} p(u^0), E_3^* p(x, 0; u^0)\},$$

element of $\vec{\mathcal{H}} = \prod_{i=1}^{3} \mathcal{H}_i$, Theorem 4.1 yields the following equivalent result:

Theorem 6.1. *Hypotheses of Theorem* 4.1. *A necessary and sufficient condition for* u^0 *to be an optimal control is that* $u^0 \in \mathcal{U}$ *satisfies*

$$(6.3)\qquad (\zeta(u^0), u - u^0)_{\vec{\mathcal{H}}} \geqq 0, \quad \forall u \in \mathcal{U},$$

where $\zeta(u^0)$ *is given by* (6.2), $p(u^0)$ *being given by* (4.8), (4.9), (4.10).

6.2 Consequence of Theorem 6.1

The set of inequalities (6.3) may also be written

$$(6.4)\qquad (\zeta(u^0), u^0)_{\vec{\mathcal{H}}} = \inf_{u \in \mathcal{U}} (\zeta(u^0), u)_{\vec{\mathcal{H}}};$$

consequently, if the set $\mathcal{U}$ is strictly convex (in the sense: every supporting hyperplane of $\mathcal{U}$ cuts $\mathcal{U}$ at a unique point) and if $\zeta(u^0) \neq 0$, the vector u^0 is uniquely determined by $\zeta(u^0)$: we have a relation of the form

$$(6.5)\qquad u^0 = -\Phi(\zeta(u^0)). \quad \square$$

Example 6.1

If $\mathcal{U}$ is the ball $\vec{\mathcal{H}}$ with radius M and center at the origin, relation (6.5) may be written (still if $\zeta(u^0) \neq 0$):

$$u^0 = -\frac{M}{\|\zeta(u^0)\|_{\vec{\mathcal{H}}}} \zeta(u^0). \quad \square$$

We can now write the *system of partial differential equations* satisfied by $y(u^0)$ and $p(u^0)$ corresponding to an optimal control u^0; thanks to (6.5), u^0 no longer appears; let us write (6.5) more explicitly:

$$(6.5\text{a})\quad \begin{cases} u_1^0 = -\Phi_1(E_1^* p(u^0), E_2^* \vec{T} p(u^0), E_3^* p(x, 0; u^0)), \\ u_2^0 = -\Phi_2(\ldots), \\ u_3^0 = -\Phi_3(\ldots). \end{cases}$$

Substituting (6.5a) into (1.1), (1.2), (1.3) and adding (4.8), (4.9), (4.10), we obtain:

Theorem 6.2. *To every optimal control u^0 (if it exists) corresponds a couple $\{y, p\}$, solution of the (nonlinear) system of partial differential equations*

$$(6.6)\qquad A y + \frac{\partial y}{\partial t} + E_1 \Phi_1(E_1^* p, E_2^* \vec{T} p, E_3^* p(x, 0)) = 0,$$

$$(6.7)\qquad \vec{B} y + E_2 \Phi_2(E_1^* p, E_2^* \vec{T} p, E_3^* p(x, 0)) = 0,$$

$$(6.8)\qquad y(x, 0) + E_3 \Phi_3(E_1^* p, E_2^* \vec{T} p, E_3^* p(x, 0)) = 0,$$

$$(6.9)\qquad A^* p - \frac{\partial p}{\partial t} - \beta_1 \Lambda_1 (y - y_d^1) = 0,$$

$$(6.10)\qquad \vec{C} p - \beta_2 \Lambda_2 (\vec{S} y - y_d^2) = 0,$$

$$(6.11)\qquad p(x, T) - \beta_3 \Lambda_3 (y(x, T) - y_d^3) = 0,$$

as long as

$$\{E_1^* p, E_2^* \vec{T} p, E_3^* p(x, 0)\} \neq 0.$$

If $\{y, p\}$ is a solution, then the corresponding optimal control u^0 is given by:

$$(6.12)\qquad u^0 = -\Phi(E_1^* p, E_2^* \vec{T} p, E_3^* p(x, 0)). \quad \square$$

Remark 6.1. For systems governed by *ordinary differential equations*, the operators Λ_1 and Λ_3 reduce to the *identity*, and the equations (6.7) and (6.10), corresponding to the boundary conditions on Σ, no longer appear. □

Remark 6.2. We have the *existence* of (at least) one optimal control, if $\mathcal{U}$ is bounded (easy verification). □

7. Consequences of the Optimality Condition (II)

7.1 Additional Hypotheses

Let:

$\mathcal{O}$ = an open set in $\mathbf{R}^N$, $\xi \in \mathcal{O}$,

$d\xi$ = Radon measure $\geqq 0$ on $\mathcal{O}$,

$\mathfrak{h}$ = Hilbert space on $\mathbf{R}$.

Assume that

(7.1) $$\vec{\mathcal{H}} = L^2(\mathcal{O}; \mathfrak{h})$$

and that

(7.2) $$\mathcal{U} = \{u \mid u \in \vec{\mathcal{H}}, u(\xi) \in \mathcal{U}_{\mathfrak{h}}, d\xi\text{-almost everywhere}\},$$

where

(7.3) $$\mathcal{U}_{\mathfrak{h}} = \text{closed convex set in } \mathfrak{h}.$$

Lemma 7.1. *The set $\mathcal{U}$ defined by* (7.2), (7.3) *is closed and convex in $\vec{\mathcal{H}}$.*

Proof. Let $u_n \in \mathcal{U}$, $u_n \to u$ in $\vec{\mathcal{H}}$. We can extract a subsequence u_ν from u_n, such that $u_\nu(\xi) \to u(\xi)$ in $\mathfrak{h}$ for almost every ξ: thus since $u_\nu(\xi)$ belongs to the set $\mathcal{U}_{\mathfrak{h}}$ which is closed in $\mathfrak{h}$, it follows that $u(\xi) \in \mathcal{U}_{\mathfrak{h}}$ a.e., therefore that $u \in \mathcal{U}$, whence the lemma. □

We shall now apply Theorem 6.1 to this situation; then $\zeta(u^0)$ is a function $\xi \to \zeta(\xi, u^0)$ of $\mathcal{O} \to \mathfrak{h}$. □

7.2 Optimality Condition

Theorem 7.1. *Under the hypotheses of Section* 7.1, *if u^0 is an optimal control, then, except possibly for ξ belonging to a set $N \subset \mathcal{O}$ of measure zero (for $d\xi$):*

(7.4) $$(\zeta(\xi; u^0), u^0(\xi))_{\mathfrak{h}} = \inf_{u \in \mathcal{U}_h} (\zeta(\xi; u^0), u)_{\mathfrak{h}},$$

ζ being defined as in Theorem 6.1, *and conversely.*

Proof. 1) Let u^0 be an optimal control. Denote by $\mathcal{L}_1$ the Lebesgue set of the (vector) function $\xi \to \zeta(\xi; u^0)$ and by $\mathcal{L}_2$ the Lebesgue set of the (scalar) function

$$\xi \to (\zeta(\xi; u^0), u^0(\xi))_{\mathfrak{h}}.$$

We shall prove (7.4) for $\xi \in \mathcal{L}_1 \cap \mathcal{L}_2$, the complement of which has measure zero.

2) Let $\xi_0 \in \mathscr{L}_1 \cap \mathscr{L}_2$ and let $\mathcal{O}_j$, $j = 1, 2, \ldots$, be a family of cubes with center ξ_0, with

$$\mathcal{O}_1 \supset \mathcal{O}_2 \supset \cdots \quad \text{and} \quad \text{measure}(\mathcal{O}_j) \to 0 \quad \text{as} \quad j \to +\infty.$$

Then thanks to the choice 1),

$$\text{(7.5)} \quad \begin{cases} \dfrac{1}{\text{measure}(\mathcal{O}_j)} \displaystyle\int_{\mathcal{O}_j} \zeta(\xi; u^0)\, d\xi \to \zeta(\xi_0; u^0) \text{ in } \mathfrak{h}, \\[2ex] \dfrac{1}{\text{measure}(\mathcal{O}_j)} \displaystyle\int_{\mathcal{O}_j} (\zeta(\xi; u^0), u^0(\xi))_{\mathfrak{h}}\, d\xi \to (\zeta(\xi_0; u^0), u^0(\xi_0))_{\mathfrak{h}}. \end{cases}$$

3) Let u be arbitrary in $\mathscr{U}_{\mathfrak{h}}$. Define

$$\text{(7.6)} \qquad u_j(\xi) = \begin{cases} u^0(\xi) & \text{outside } \mathcal{O}_j, \\ u & \text{in } \mathcal{O}_j. \end{cases}$$

In this manner we define an "admissible control", i.e. $u_j \in \mathscr{U}$ (see (7.2)) and therefore, according to Theorem 6.1:

$$\text{(7.7)} \qquad (\zeta(u^0), u_j - u^0)_{\mathscr{X}} \geqq 0,$$

or

$$\text{(7.8)} \qquad \int_{\mathcal{O}_j} (\zeta(\xi; u^0), u - u^0(\xi))_{\mathfrak{h}}\, d\xi \geqq 0.$$

Dividing by measure $(\mathcal{O}_j)$, it follows that

$$\frac{1}{\text{measure}(\mathcal{O}_j)} \int_{\mathcal{O}_j} (\zeta(\xi; u^0), u)_{\mathfrak{h}}\, d\xi - \\ - \frac{1}{\text{measure}(\mathcal{O}_j)} \int_{\mathcal{O}_j} (\zeta(\xi; u^0), u^0(\xi))_{\mathfrak{h}}\, d\xi \geqq 0$$

and as $j \to \infty$, we obtain, with (7.5):

$$\text{(7.9)} \qquad (\zeta(\xi_0; u^0), u)_{\mathfrak{h}} \geqq (\zeta(\xi_0; u^0), u^0(\xi_0))_{\mathfrak{h}},$$

whence (7.4).

Conversely, if we have (7.4), then we have (7.9), or

$$(\zeta(\xi; u^0), u(\xi))_{\mathfrak{h}} \geqq (\zeta(\xi; u^0), u^0(\xi))$$

for almost all ξ, if $u \in \mathscr{U}$, and therefore we have (6.3) (after integration in ξ over $\mathcal{O}$). □

Remark 7.1. If $\mathfrak{h} = \prod_{i=1}^{i_0} L^2(\mathcal{O}^i; \mathfrak{h}_i)$, $\mathcal{O}^i$ = open sets in $\mathbf{R}^{N_i}$ provided with the Lebesgue measure, and if

$$\mathscr{U}_{\mathfrak{h}} = \prod_{i=1}^{i} \mathscr{U}_{\mathfrak{h}_i}, \tag{7.10}$$

where

$$\mathscr{U}_{\mathfrak{h}_i} = \text{closed convex set in } \mathfrak{h}_i, \tag{7.11}$$

then (7.4) *"decomposes" into* i_0 *conditions* $(1 \leqq i \leqq i_0)$:

$$\begin{cases} (\zeta_i(\xi^i; u^0), u_i^0(\xi^i))_{\mathfrak{h}_i} = \inf\limits_{u_i \in \mathscr{U}_{\mathfrak{h}_i}} (\zeta_i(\xi^i; u^0), u_i)_{\mathfrak{h}_i}, \\ \text{except possibly for } \xi^i \text{ in a set of measure zero on } \mathcal{O}^i. \end{cases} \tag{7.12}$$ □

8. Complements on the Choice of the Spaces K_i

8.1 Orientation

The nonlinear system of partial differential equations (6.6), . . ., (6.12) contains the isomorphisms Λ_i, which we must now specify completely. We recall (Definition 4.1) that Λ_i is the canonical isomorphism of K_i onto K_i'.

In the examples, the K_i's are (integer or fractional order) Sobolev spaces and the Λ_i's are therefore differential or integro-differential operators, or Green operators, according to whether the order of K_i is >0 or <0 (if K_i has order 0, therefore is an L^2-space, Λ_i is the identity).

8.2 Choice of $K_1(Q)$

The *simplest* choice for $K_1(Q)$ (but this choice *is not* mandatory and for this reason we have preserved the general notation $K_1(Q)$) is

$$K_1(Q) = L^2(Q) \tag{8.1}$$

and thus

$$\Lambda_1 = \text{identity}. \quad \square \tag{8.2}$$

8.3 Choice of $K_2(\Sigma)$

We seek $K_2(\Sigma)$ in the form

$$K_2(\Sigma) = \prod_{j=0}^{m-1} H^{\alpha_j}(\Sigma), \quad \text{integer } \alpha_j. \tag{8.3}$$

If

$$\mu - \mu_j - \tfrac{1}{2} = (\min m_j) - \mu_j \geqq 0, \tag{8.4}$$

we take

(8.5) $H^{\alpha_j}(\Sigma) = L^2(\Sigma)$ (and then the corresponding operator Λ_{2j} is the identity).

If

$$\mu - \mu_j - \tfrac{1}{2} < 0, \tag{8.6}$$

we take

$$\alpha_j = (\min m_j) - \mu_j \qquad (<0). \tag{8.7}$$

In this case Λ_{2j} is given by

$$\begin{cases} \Lambda_{2j}\,\omega = (-\Delta_\Sigma + I)^{\alpha_j}\,\omega \text{ ((1))}, \text{ with Dirichlet} \\ \text{conditions on } \Sigma,\ \omega \in H^{\alpha_j}(\Sigma). \quad \square \end{cases} \tag{8.8}$$

8.4 Choice of $K_3(\Omega)$

We may take

$$K_3(\Omega) = H^{\min m_j - m}(\Omega). \tag{8.9}$$

Therefore

$$\Lambda_3 = \text{identity}, \quad \text{if } \min m_j = m, \tag{8.9$_1$}$$

and

$$\begin{cases} \Lambda_3\,\omega = (-\Delta + I)^{\min m_j - m}\,\omega, \quad \text{with Dirichlet boundary} \\ \text{conditions, if } \min m_j < m,\ \omega \in K_3(\Omega). \quad \square \end{cases} \tag{8.9$_2$}$$

9. Examples

Example 9.1

Consider the following very particular case of (1.1), (1.2), (1.3):

$$\begin{cases} -\Delta y + \dfrac{\partial y}{\partial t} = 0, \\ y\,|_\Sigma = u_2 = u, \\ y(x, 0) = 0. \end{cases} \tag{9.1}$$

(Therefore $E_1 = E_3 = 0$, $\mathscr{H}_2 = L^2(\Sigma)$, $E_2 =$ identity.)

Then $m = 1$, $m_0 = 0$ and, according to Section 8.3, we take

$$K_2(\Sigma) = H^{-1}(\Sigma).$$

((1)) $\Delta_\Sigma =$ Laplace-Beltrami operator on Σ.

Thus the adjoint system (4.8), (4.9), (4.10) is

$$-\Delta p - \frac{\partial p}{\partial t} = 0, \tag{9.2}$$

$$(-\Delta_\Sigma + I)(p|_\Sigma) = \beta_2 \left(\frac{\partial}{\partial \nu} y \bigg|_\Sigma - y_d^2 \right), \tag{9.3}$$

$$p|_\Sigma = 0 \quad \text{for} \quad t = 0 \quad \text{and} \quad t = T, \tag{9.4}$$

$$p(x, T) = 0. \tag{9.5}$$

[Equations (9.3), (9.4) *define* $p|_\Sigma$, which, together with (9.2) and (9.5), determines p in Q]. □

Example 9.2

Now consider the system

$$\begin{cases} -\Delta y + \dfrac{\partial y}{\partial t} = 0, \\ \dfrac{\partial y}{\partial \nu}\bigg|_\Sigma = u_2 = u, \\ y(x, 0) = 0. \end{cases} \tag{9.6}$$

Therefore $E_1 = E_3 = 0$, $\mathscr{H}_2 = L^2(\Sigma)$, $E_2 =$ identity. Then we have $\mu_0 = 0$, $m = 1$, $m_0 = 1$ and therefore

$$\min m_j - \mu_0 = 1$$

and therefore (see (8.4), (8.5)) we take

$$K_2(\Sigma) = L^2(\Sigma).$$

Thus the adjoint system is

$$-\Delta p - \frac{\partial p}{\partial t} = 0, \tag{9.7}$$

$$\frac{\partial p}{\partial \nu}\bigg|_\Sigma = \beta_2 (y|_\Sigma - y_d^2), \tag{9.8}$$

$$p(x, T) = 0. \quad \square \tag{9.9}$$

Remark 9.1. For the case in point, we could have chosen

$$K_2(\Sigma) = H^{1,1/2}(\Sigma).$$

Then Λ_2 is an integro-differential operator. System (9.7), (9.8), (9.9) is replaced by

$$(9.10)\qquad \begin{cases} -\Delta p - \dfrac{\partial p}{\partial t} = 0, \\ \dfrac{\partial p}{\partial \nu}\Big|_{\Sigma} = \beta_2 \Lambda_2 (y|_{\Sigma} - y_d^2), \\ p(x, T) = 0. \end{cases}$$

The cost function, *in the first case* of system (9.7), (9.8), (9.9), is

$$(9.11)\qquad \beta_2 \int_{\Sigma} (y(x, t; u) - y_d^2)^2 \, d\Sigma$$

and *in the second case* (of (9.10)) is

$$(9.12)\qquad \beta_2 \| y(x, t; u) - y_d^2 \|^2_{H^{1,1/2}(\Sigma)}.$$

Thus, if we seek an optimal approximation in a larger norm (i.e. a finer topology), the adjoint problem is more complicated (as one would expect). □

Example 9.3

Consider

$$(9.13)\qquad \begin{cases} -\Delta y + \dfrac{\partial y}{\partial t} = 0, \\ \dfrac{\partial y}{\partial \nu}\Big|_{\Sigma} = 0, \\ y(x, 0) = u_3 = u. \end{cases}$$

Then $E_1 = E_2 = 0$, $\mathscr{H}_3 = L^2(\Omega)$, $E_3 =$ identity. We have

$$m = 1, \qquad m_0 = 1,$$

and therefore we have the particular case of Section 5.

Thus we take (see (5.5)) $K_3(\Omega) = L^2(\Omega)$ and obtain the adjoint system:

$$(9.14)\qquad \begin{cases} -\Delta p - \dfrac{\partial p}{\partial t} = 0, \\ \dfrac{\partial p}{\partial \nu}\Big|_{\Sigma} = 0, \\ p(x, T) = \beta_3 (y(x, T) - y_d^3). \quad □ \end{cases}$$

Example 9.4

Consider

$$(9.15)\qquad \begin{cases} -\Delta y + \dfrac{\partial y}{\partial t} = 0, \\ y|_{\Sigma} = 0, \\ y(x,0) = u_3 = u. \end{cases}$$

Therefore $E_1 = E_2 = 0$, $\mathscr{H}_3 = L^2(\Omega)$, $E_3 =$ identity. We have $m = 1$, $m_0 = 0$ and therefore (see (8.9)):

$$K_3(\Omega) = H^{-1}(\Omega).$$

Then the adjoint system is

$$(9.16)\qquad -\Delta p - \frac{\partial p}{\partial t} = 0,$$

$$(9.17)\qquad p|_{\Sigma} = 0,$$

$$(9.18)\qquad \begin{cases} (-\Delta + \mathrm{I})\, p(x,T) = \beta_3(y(x,T) - y_d^3), \\ p(x,T) = 0 \quad \text{for} \quad x \in \Gamma. \end{cases}$$

In this case, (9.18) determines $p(x,T)$ (Dirichlet problem!), which together with (9.16), (9.17) defines p in Q. □

Example 9.5

Finally, we consider an example with $m > 1$. Let the system be

$$(9.19)\qquad \Delta^2 y + \frac{\partial y}{\partial t} = 0,$$

$$(9.20)\qquad \begin{cases} y|_{\Sigma} = u_{21}, \\ \left.\dfrac{\partial y}{\partial \nu}\right|_{\Sigma} = u_{22}, \end{cases}$$

$$(9.21)\qquad y(x,0) = 0.$$

Then $E_1 = E_3 = 0$, $\mathscr{H}_2 = (L^2(\Sigma))^2$, $E_2 =$ identity.

We have: $m = 2$, $m_0 = 0$, $m_1 = 1$, $\mu_0 = 3$, $\mu_1 = 2$, $\min m_j = 0$ and, according to (8.4) and (8.5), we therefore take:

$$(9.22)\qquad K_2(\Sigma) = H^{-3}(\Sigma) \times H^{-2}(\Sigma),$$

and the adjoint system becomes:

(9.23)
$$\Delta^2 p - \frac{\partial p}{\partial t} = 0,$$

(9.24)
$$\begin{cases} (-\Delta_\Sigma + I)^3 \, p\,|_\Sigma = \beta_2 \left(\left.\dfrac{\partial \Delta y}{\partial \nu}\right|_\Sigma - y_{d_1}^2 \right), \\ p\,|_\Sigma = \left.\dfrac{\partial p}{\partial t}\right|_\Sigma = \left.\dfrac{\partial^2 p}{\partial t^2}\right|_\Sigma = 0, \quad \text{for } t = 0 \text{ and } t = T, \end{cases}$$

(9.25)
$$\begin{cases} (-\Delta_\Sigma + I)^2 \left.\dfrac{\partial p}{\partial \nu}\right|_\Sigma = \beta_2 (-\Delta y\,|_\Sigma - y_{d_2}^2), \\ \left.\dfrac{\partial p}{\partial \nu}\right|_\Sigma = \dfrac{\partial}{\partial t}\left(\left.\dfrac{\partial p}{\partial \nu}\right|_\Sigma \right) = 0, \quad \text{for } t = 0 \text{ and } t = T \end{cases}$$

(9.26)
$$p(x, T) = 0.$$

Equations (9.24) and (9.25) define $p|_\Sigma$ and $\left.\frac{\partial p}{\partial \nu}\right|_\Sigma$, which together with (9.23) and (9.26) defines p in Q. □

10. Non-Parabolic Cases. Statement of the Problems. Generalities

Here, we introduce the framework of the theory in a somewhat different axiomatic fashion, in order to insist on the numerous possible variants of control problems, when the controls appear in the boundary conditions, and because of the inherent difficulties of non-homogeneous, non-parabolic problems noted in Chapter 5.

10.1 Notation

For a control

$$u = \{u_1, u_2, u_3, u_4\} \in \vec{\mathscr{H}} = \prod_{i=1}^{4} \mathscr{H}_i,$$

$\mathscr{H}_i$ a Hilbert space on $\mathbf{R}$, we consider a system for which the state y is given by

(10.1)
$$A\,y + y'' = E_1 u_1,$$

(10.2)
$$\vec{B}\,y = E_2 u_2,$$

(10.3)
$$y(x, 0) = E_3 u_3,$$

(10.4)
$$y'(x, 0) = E_4 u_4,$$

where:

the operators A, $\{B_j\} = \vec{B}$ satisfy the conditions of Chapter 5, Section 1, and the operators E_i satisfy

$$E_i \in \mathscr{L}(\mathscr{H}_i, \mathscr{L}_i), \tag{10.5}$$

where:

$\mathscr{L}_1 = \mathscr{L}_1(Q)$ is a space of functions or distributions or "functionals" on Q,

$\mathscr{L}_2 = \mathscr{L}_2(\Sigma)$ is a space of functions or distributions on Σ,

$\mathscr{L}_3 = \mathscr{L}_3(\Omega)$, $\mathscr{L}_4 = \mathscr{L}_4(\Omega)$ are spaces of functions or distributions on Ω.

We assume the spaces $\mathscr{L}_i$ to be *chosen so that* (10.1), ..., (10.4) *admits a unique solution, either the "usual" solution, or the solution defined by transposition*, i.e. (see Chapter 5):

$$(y, A^* \Phi + \Phi'')_Q = (E_1 u_1, \Phi)_Q + (E_2 u_2, \vec{T} \Phi)_\Sigma + \\ + (E_4 u_4, \Phi(0))_\Omega - (E_3 u_3, \Phi(0))_\Omega \tag{10.6}$$

for every Φ such that

$$\begin{cases} A^* \Phi + \Phi = \psi, \\ \vec{B} \Phi = 0, \\ \Phi(T) = 0, \quad \Phi'(T) = 0, \end{cases} \tag{10.7}$$

where ψ describes the *dual* of the space described by y. □

Remark 10.1. Even in practical examples, we have some freedom in the choice of the $\mathscr{L}_i$'s. □

The solution of (10.1), ..., (10.4) (possibly in the sense of (10.6)) is denoted by

$$y = y(u) = g(x, t; u). \tag{10.8}$$

10.2 Cost Function

When $\{E_1 u_1, E_2 u_2, E_3 u_3, E_4 u_4\} \in \prod_{i=1}^{4} \mathscr{L}_i$, the solution $y(u)$ belongs *to a space* $\mathscr{Y}$ (the complete characterization of which may be delicate; see Chapter 5). □

Formally, the *control problem* is the following: *as u describes a set $\mathscr{U} \subset \prod_{i=1}^{4} \mathscr{H}_i$, minimize the "distance" from $y(u)$ to y_d, with y_d given in $\mathscr{Y}$.*

We shall always assume that the $\mathscr{L}_i$'s *are Hilbert spaces*, and therefore so is $\mathscr{Y}$. Thus we want to minimize $\|y(u) - y_d\|_{\mathscr{Y}}$. □

As for the parabolic case (see Section 1), it is useful to consider, not only y on Q, but also $\vec{S}\, y$ (on Σ) and $y(T;u)$ and $y'(T;u)$.

We introduce the following spaces (compare with Sections 2 and 8):

$$K_1(Q) \supset \mathcal{Y} \quad \text{(with the possibility of equality)}, \tag{10.9}$$

$$K_2(\Sigma) \quad \text{such that} \quad \vec{S} \in \mathcal{L}(\mathcal{Y}; K_2(\Sigma)), \tag{10.10}$$

$$\begin{cases} K_3(\Omega),\ K_4(\Omega) \ \text{ such that } \ y(u) \to \{y(T;u), y'(T;u)\} \\ \text{is a continuous mapping of } \ \mathcal{Y} \to K_3(\Omega) \times K_4(\Omega). \end{cases} \tag{10.11}$$

We now introduce the *cost function*:

$$\begin{cases} J(u) = \beta_1 \|y(u) - y_d^1\|_{K_1}^2 + \beta_2 \|\vec{S}\, y(u) - y_d^2\|_{K_2}^2 + \\ \qquad + \beta_3 \|y(T;u) - y_d^3\|_{K_3}^2 + \beta_4 \|y'(T;u) - y_d^4\|_{K_4}^2, \\ \beta_i \geqq 0, \quad \sum_{i=1}^{4} \beta_i > 0, \end{cases} \tag{10.12}$$

where $y_d = \{y_d^1, y_d^2, y_d^3, y_d^4\}$ is given in $\prod_{i=1}^{4} K_i$. □

The control problem now is:

$$\begin{cases} \text{let } \mathcal{U} \text{ be a closed convex set in } \prod_{i=1}^{4} \mathcal{H}_i; \text{ find} \\ \inf_{u \in \mathcal{U}} J(u). \end{cases} \tag{10.13}$$

If $u^0 \in \mathcal{U}$ satisfies

$$J(u^0) \leqq J(u), \quad \forall u \in \mathcal{U}, \tag{10.14}$$

then u^0 *is called an optimal control.* □

Remark 10.2. The analogue to Remark 1.2, Section 1 holds. □

10.3 Optimality Condition (I)

The remarks leading to Proposition 4.1 are evidently valid here and yield:

Proposition 10.1 *A necessary and sufficient condition for u^0 to be an optimal control is that:*

$$\begin{cases} \beta_1 (y(u^0) - y_d^1, y(u) - y(u^0))_{K_1} + \\ + \beta_2 (\vec{S}\, y(u^0) - y_d^2, \vec{S}\, y(u) - \vec{S}\, y(u^0))_{K_2} + \\ + \beta_3 (y(T;u^0) - y_d^3, y(T;u) - y(T;u^0))_{K_3} + \\ + \beta_4 (y'(T;u^0) - y_d^4, y'(T;u) - y'(T;u^0))_{K_4} \geqq 0 \quad \forall u \in \mathcal{U}. \end{cases} \tag{10.15}$$

10.4 Adjoint Problem

As in Definition 4.1 of Section 4.2, we define Λ_i to be the canonical isomorphism of K_i onto K_i'.

Next, we define $p(x, t; u) = p(u)$ to be the solution of (the adjoint problem):

$$A^* p(u) + p''(u) = \beta_1 \Lambda_1 (y(u) - y_d^1), \tag{10.16}$$

$$\vec{C} p(u) = \beta_2 \Lambda_2 (\vec{S} y(u) - y_d^2), \tag{10.17}$$

$$p(x, T; u) = \beta_3 \Lambda_3 (y(x, T; u) - y_d^3), \tag{10.18}$$

$$p'(x, T; u) = -\beta_4 \Lambda_4 (y'(x, T; u) - y_d^4). \quad \square \tag{10.19}$$

Let us specify *in what sense* $p(u)$ is a solution of (10.16), ..., (10.19). Consider ω, solution of

$$\begin{cases} A\,\omega + \omega'' = \chi, & \chi \text{ given in } \mathscr{L}_1(Q), \\ \vec{B}\,\omega = 0, & \\ \omega(x, 0) = 0, & \omega'(x, 0) = 0. \end{cases} \tag{10.20}$$

Then (by definition) $p(u)$ satisfies (10.16), ..., (10.19) if

$$\begin{cases} (p(u), A\,\omega + \omega'') = \beta_1 \langle \Lambda_1 (y(u) - y_d^1), \omega \rangle + \\ \qquad + \beta_2 \langle \vec{S}\, y(u) - y_d^2, \vec{S}\,\omega \rangle - \\ \qquad - \beta_3 \langle y(T; u) - y_d^3, \omega'(T) \rangle - \\ \qquad - \beta_4 \langle y'(T; u) - y_d^4, \omega(T) \rangle. \end{cases} \tag{10.21}$$

More precisely:

Proposition 10.2. *There exists a unique* $p(u) \in \mathscr{L}_1'(Q)$, *satisfying* (10.21) $\forall \omega$ *given by* (10.20) (*for arbitrary* $\chi \in \mathscr{L}_1(Q)$).

Proof. For $\chi \in \mathscr{L}_1(Q)$, problem (10.20) admits a unique solution in $\mathscr{Y}$ (by definition of $\mathscr{Y}$; see Section 10.2). Then (see (10.9), ..., (10.11))

$$\omega \to \{\omega, \vec{S}\,\omega, \omega(T), \omega'(T)\} \quad \text{is a continuous mapping of } \mathscr{Y} \to \prod_{i=1}^{4} K_i,$$

and therefore the right-hand side of (10.21) is a linear form, say $\omega \to M(\omega)$, which is *continuous on* $\mathscr{Y}$.

If $\mathscr{Y}_0$ ($\subset \mathscr{Y}$) is the space described by ω as χ describes $\mathscr{L}_1$, the operator $A + \dfrac{\partial^2}{\partial t^2}$ is an isomorphism of $\mathscr{Y}_0$ onto $\mathscr{L}_1$ and therefore, by transposition (see Chapter 5), there exists a unique $p(u) \in \mathscr{L}_1'(Q)$ satisfying (10.21). $\square$

10.5 Green's Formula

We now make the *hypothesis*:

$$(10.22)\quad \begin{cases} \text{for } \{l_2, l_3, l_4\} \in \mathscr{L}_2 \times \mathscr{L}_3 \times \mathscr{L}_4, \text{ there exists } \chi \in K_1(Q) \\ \text{such that} \\ A\chi + \chi'' \in \mathscr{L}_1, \\ \vec{B}\chi = l_2, \quad \chi(0) = l_3, \quad \chi'(0) = l_4, \\ \vec{S}\chi = 0, \quad \chi(T) = 0, \quad \chi'(T) = 0, \\ \text{the mapping } \{l_2, l_3, l_4\} \to \chi \text{ being linear and continuous.} \end{cases}$$

Then, if $p \in \mathscr{L}_1'$ with $A^* p + p'' \in K_1'$, we can *define*

$$(10.23)\quad \begin{cases} \vec{T} p \in \mathscr{L}_2', \\ p(0) \in \mathscr{L}_4', \\ p'(0) \in \mathscr{L}_3', \end{cases}$$

with the help of

$$(10.24)\quad \begin{cases} \langle p, A\chi + \chi''\rangle - \langle A^* p + p'', \chi\rangle \\ \qquad = -\langle \vec{T} p, l_2\rangle - \langle p(0), l_4\rangle + \langle p'(0), l_3\rangle. \end{cases}$$

Remark 10.3. In fact, for each particular case, we shall need to demonstrate (as we have done in Chapter 4 and 5) that the mappings $p \to \vec{T} p, p(0), p'(0)$ may be defined by *extension by continuity*. Furthermore, the property of belonging to $\mathscr{L}_2 \times \mathscr{L}_3 \times \mathscr{L}_4$ *may* contain *compatibility relations* (see Chapter 4 and 5), so that we cannot "separate" the "components" $\vec{T} p$, $p(0)$, $p'(0)$ in (10.23): $\{\vec{T} p, p(0), p'(0)\}$ belongs to a distribution space *quotient*. □

Under hypothesis (10.22) (and within the restrictions of Remark 10.3) we have (done what was necessary to have) *Green's formula*:

Proposition 10.3. *For u and u^0 given in $\vec{\mathscr{H}}$:*

$$(10.25)\quad \begin{cases} \langle A y(u) + y''(u), p(u^0)\rangle - \langle y(u), A^* p(u^0) + p''(u^0)\rangle \\ \qquad = \langle \vec{S} y(u), \vec{C} p(u^0)\rangle - \langle \vec{B} y(u), \vec{T} p(u^0)\rangle + \\ \qquad + \langle y'(T; u), p(T; u^0)\rangle - \langle y'(0; u), p(0; u^0)\rangle - \\ \qquad - \langle y(T; u), p'(T; u^0)\rangle + \langle y(0; u), p'(0; u^0)\rangle, \end{cases}$$

where on the left-hand (resp. right-hand) side of the equation the brackets denote, in order, the duality between $\mathscr{L}_1$ and $\mathscr{L}_1'$, K_1 and K_1' (resp. K_2, K_2'; $\mathscr{L}_2, \mathscr{L}_2'$; K_4, K_4'; $\mathscr{L}_4, \mathscr{L}_4'$; K_3, K_3'; $\mathscr{L}_3, \mathscr{L}_3'$). □

10.6 Optimality Condition (II)

Theorem 10.1 *Hypotheses of Section* 10.1 *and* (10.22). *A necessary and sufficient condition for u^0 to be an optimal control is that $u^0 \in \mathcal{U}$ satisfies*

$$(10.26)\quad \begin{cases} \langle p(u^0), E_1(u_1 - u_1^0)\rangle + \langle \vec{T}\, p(u^0), E_2(u_2 - u_2^0)\rangle + \\ \qquad + \langle -p'(0; u^0), E_3(u_3 - u_3^0)\rangle + \\ \qquad + \langle p(0; u^0), E_4(u_4 - u_4^0)\rangle \geqq 0 \quad \forall u \in \mathcal{U}, \end{cases}$$

where $p(u^0)$ is the solution of (10.16), ..., (10.19) (in the sense of (10.21)) (with $u = u^0$).

Proof. In (10.25), substitute $u - u^0$ for u; the equality of the left-hand sides of (10.15) and (10.26) follows, whence the theorem. □

10.7 Consequences

Results analogous to Sections 5 and 6 follow from Theorem 10.1. □

11. Applications. Examples

11.1 Control in the Boundary Conditions

From now on, we shall assume that the control appears only in the boundary conditions.

We take

$$(11.1)\quad \begin{cases} E_1 = 0, & E_3 = 0, & E_4 = 0, \\ \mathcal{H}_2 = (L^2(\Sigma))^m, & \mathcal{L}_2 = (L^2(\Sigma))^m, & E_2 = \text{identity}. \end{cases}$$

Thus, setting $u_2 = u$, the state y of the system is defined by

$$(11.2)\quad \begin{cases} A\, y + y'' = 0, \\ \vec{B}\, y = u \in (L^2(\Sigma))^m, \\ y(0) = 0, \\ y'(0) = 0. \end{cases}$$

We take the cost function:

$$(11.3)\qquad J(u) = \beta_1 \| y(u) - y_d \|_{K_1}^2 + \beta_2 \| \vec{S}\, y(u) - y_d^2 \|_{K_2}^2$$

and we now have to specify how to choose K_1 and K_2.

11.2 Choice of K_1

We apply Theorem 6.1, Chapter 5 (for the case of Dirichlet boundary conditions, we can also apply Theorem 6.2, Chapter 5; see also Remark 5.3, Chapter 5).

Thus, let

$$\theta = 1 - \frac{1}{6m} - \frac{1}{3m} \min m_j . \tag{11.4}$$

Then

$$y \in H^{-\theta}(0, T; H^{2m(1-\theta)}(\Omega)) \cap H^{2-3\theta}(0, T; H^0(\Omega)) = \mathcal{Y}, \tag{11.5}$$

except for the two following particular cases:

$$\begin{cases} m = 1, \quad m_0 = 1; \quad \text{then} \\ y \in H_{00}^{-1/2}(0, T; H^1(\Omega)) \cap H^{1/2}(0, T; H^0(\Omega)) = \mathcal{Y} \end{cases} \tag{11.6}$$

and

$$\begin{cases} m\text{-odd}, \quad \min m_j = \dfrac{m-1}{2}, \\ y \in H^{-5/6}(0, T; H^{m/3}(\Omega)) \cap H_{00}^{-1/2}(0, T; H^0(\Omega)) = \mathcal{Y}. \end{cases} \tag{11.7}$$

It is required to choose $K_1 = K_1(\Omega)$, a "simple" space such that

$$\mathcal{Y} \subset K_1 .$$

We distinguish two cases, according to whether $2 - 3\theta \geqq 0$ or <0.

First case:

$2 - 3\theta \geqq 0$, θ defined by (11.4), is equivalent to

$$\min m_j \geqq m - \tfrac{1}{2},$$

that is

$$\min m_j = m \tag{11.8}$$

(note that this also corresponds to (11.6)).

Then in any case (i.e. (11.5) and (11.6)) we have: $\mathcal{Y} \subset L^2(Q)$ and therefore:

Proposition 11.1. *If* (11.8) *holds, we may take*

$$K_1 = L^2(Q). \quad \square \tag{11.9}$$

Second case.

Assume that $2 - 3\theta < 0$. We still have

$$2 - 3\theta > -1$$

and in any case (i.e. (11.5) and (11.7)) we have

$$\mathcal{Y} \subset H^{-1}(0, T; H^0(\Omega)).$$

Then:

Proposition 11.2. *If* (11.8) *does not hold, we may take*

(11.10) $$K_1 = H^{-1}(0, T; H^0(\Omega)). \quad \square$$

Remark 11.1. For the case (11.10) we have:

$$K_1' = H_0^1(0, T; H^0(\Omega));$$

if we provide K_1' with the scalar product

$$\int_0^T [(u, v)_{H^0(\Omega)} + (u', v')_{H^0(\Omega)}]\, dt,$$

the operator Λ_1 of the general theory is given in the following way:

let f be given in $H^{-1}(0, T; H^0(\Omega))$; let ω be the solution of

(11.11) $$\begin{cases} -\omega'' + \omega = f \\ \omega(0) = 0, \quad \omega(T) = 0. \end{cases}$$

Then

$$\omega = \Lambda_1 f. \quad \square$$

11.3 Choice of K_2

Now, the problem is to "characterize" the space described by $\tilde{S} y$ as y describes $\mathscr{Y}$ and to choose $K_2 = K_2(\Sigma)$, a "simple" space such that

$$\tilde{S}\, \mathscr{Y} \subset K_2. \quad \square$$

The result (11.5) does not apply to all cases; indeed for the present case (A having symmetric principal part) we have:

$$\text{order } \tilde{S}_j = 2m - m_j - 1$$

and we cannot define $\tilde{S}_j y$ *by using only* (11.5), unless

$$2m(1-\theta) - \tfrac{1}{2} > 2m - m_j - 1, \quad \theta \text{ given by (11.4)},$$

i.e. unless

$$m_j \geq 2m - \tfrac{2}{3} \min m_j + \tfrac{1}{6},$$

which is not always realized for all j (and may not be realized for *any* j). $\square$

We shall therefore reconsider the question directly by interpolation, using the results of Chapter 5.

First of all, Theorem 3.1, Chapter 5 yields:

if $\quad u_j \in H_0^1(0, T; H^{2m-m_j-1/2}(\Gamma)) \cap H_0^{(3m-(m_j+1/2))/m}(0, T; H^0(\Gamma)),$

to which we must add the global relations $\frac{1}{\sqrt{t}} u'_j \in L^2(0, T; H^0(\Gamma))$ in the exceptional case:

$$m\text{-odd} \quad \text{and} \quad m_j = \frac{3m-1}{2};$$

then $y \in H^{2m, 2}_{,0}(Q)$ and therefore

$$\vec{S}\, y \in \prod_{j=0}^{m-1} H^{2m-(2m-m_j-1)-1/2, (2m-(2m-m_j-1)-1/2)/m}_{,0}(\Sigma)$$

from which it follows that:

$$(11.12) \begin{cases} u \to \vec{S}\, y \quad \text{is a continuous linear mapping of} \\ \prod_{j=0}^{m-1} H_0^1(0, T; H^{2m-m_j-1/2}(\Gamma)) \cap H_0^{(3m-m_j-1/2)/m}(0, T; H^0(\Gamma)) \to \\ \qquad \to \prod_{j=0}^{m-1} H^{m_j+1/2, (m_j+1/2)/m}_{,0}(\Sigma). \quad \square \end{cases}$$

Let us now apply Theorem 4.1, Chapter 5. We see that if

$$u \in \prod_{j=0}^{m-1} (H^{m_j+1/2, (m_j+1/2)/m}_{,0}(\Sigma))', \quad \text{then} \quad y \in H^{0,-1}(Q).$$

In the present case, we have: $A\, y + y'' = 0$, therefore in particular

$$A\, y + y'' \in L^2(Q).$$

For the moment, we denote by F the space:

$$F = \{y \mid y \in H^{0,-1}(Q),\, A\, y + y'' \in L^2(Q)\}$$

and (see Chapter 5, Section 11 for a similar situation)

$$\mathbf{F} = \text{closure of } \mathscr{D}(\bar{Q}) \text{ in } F.$$

(we do not know whether $\mathbf{F}$ is really *distinct* from F). Then $y \in \mathbf{F}$ and we shall verify that

$$(11.13) \begin{cases} \text{for } y \in \mathbf{F}, \quad \text{we have} \\ \vec{S}\, y \in \prod_{j=0}^{m-1} (H_0^1(0, T; H^{2m-m_j-1/2}(\Gamma)) \cap \\ \cap H_0^{(3m-m_j-1/2)/m}(0, T; H^0(\Gamma)))', \\ \text{the mapping } y \to \vec{S}\, y \text{ being continuous for the natural} \\ \text{topologies.} \end{cases}$$

Indeed, we use Green's formula:

$$(A\, y + y'', \omega) - (y, A^* \omega + \omega'') = \langle \vec{S}\, y, \overline{\vec{C}\, \omega} \rangle_\Sigma - \langle \vec{B}\, y, \overline{\vec{T}\, \omega} \rangle_\Sigma,$$

where we must take (so that the scalar products on the left-hand side of the equation are well-defined):

$$(11.14)\quad \begin{cases} \omega \in L^2(0, T; H^{2m}(\Omega)), \quad \omega' \in H^{2m,2}(Q), \\ \omega(x, 0) = \omega'(x, 0) = \omega''(x, 0) = \omega(x, T) \\ \qquad = \omega'(x, T) = \omega''(x, T) = 0. \end{cases}$$

Then, if $\vec{\psi}$ is given in the space

$$(11.15)\ \mathscr{G}_1 = \prod_{j=0}^{m-1} H_0^1(0, T; H^{2m-m_j-1/2}(\Gamma)) \cap H_0^{(3m-m_j-1/2)/m}(0, T; H^0(\Gamma))$$

(with possibly the global relations which we already noted), we can find (see Chapter 5, Proposition 3.3) ω which satisfies (11.14) and $\vec{C}\,\omega = \vec{\psi}$ and $\vec{T}\,\omega = 0$, and which depends continuously on $\vec{\psi}$. (11.13) follows. □

Thus, it follows from (11.12) and (11.13) that

$$(11.16)\quad \begin{cases} u \to \vec{S}\, y \quad \text{is a continuous linear mapping of} \\ \mathscr{G}_1 \to \mathscr{G}_2 = \prod_{j=0}^{m-1} H_{,0}^{m_j+1/2,(m_j+1/2)/m}(\Sigma) \\ \text{and of} \\ \mathscr{G}_2' \to \mathscr{G}_1'. \end{cases}$$

By interpolation we obtain:

Proposition 11.3. *Let* $\mathscr{G}_1$ *and* $\mathscr{G}_2$ *be defined as in* (11.5) *and* (11.16). *Then, for* $0 < \theta < 1$,

$$u \to \vec{S}\, y, \quad y \text{ solution of } (11.2),$$

is a continuous linear mapping of

$$(11.17)\quad [\mathscr{G}_1, \mathscr{G}_2']_\theta \to [\mathscr{G}_2, \mathscr{G}_1']_\theta. \quad \square$$

Remark 11.2. We recall that (duality theorem, Chapter 1, Section 6.2)

$$(11.18)\quad [\mathscr{G}_2, \mathscr{G}_1']_\theta = [\mathscr{G}_1, \mathscr{G}_2']_{1-\theta}'. \quad \square$$

Now we choose θ such that

$$(11.19)\quad L^2(\Sigma)^m \subset [\mathscr{G}_1, \mathscr{G}_2']_\theta.$$

But

$$\mathscr{G}_1 \supset \prod_{j=0}^{m-1} H_0^{3m-m_j-1/2,(3m-m_j-1/2)/m}(\Sigma)$$

(with the global relations already noted) and therefore

(11.20)

$$[\mathcal{G}_1, \mathcal{G}_2']_\theta \supset \prod_{j=0}^{m-1} [H_{,0}^{3m-m_j-1/2,(3m-m_j-1/2)/m}(\Sigma), (H_{,0}^{m_j+1/2,(m_j+1/2)/m}(\Sigma))']_\theta,$$

a space which, by Chapter 1, Section 12, contains $L^2(\Sigma)^m$ if

$$3m(1-\theta) - m_j - \tfrac{1}{2} \leqq 0, \quad \forall j,$$

i.e. if

$$\theta = 1 - \frac{1}{6m} - \frac{1}{3m} \min m_j;$$

we recover the θ given by (11.4).

Consequently:

$$\text{(11.21)} \quad \begin{cases} \text{if } u \in L^2(\Sigma)^m, \text{ then } \vec{S}y \in [\mathcal{G}_2, \mathcal{G}_1']_\theta, \\ \theta \text{ given by (11.4).} \end{cases}$$

But, by duality, it follows from (11.20) that

$$\text{(11.22)} \quad \begin{cases} [\mathcal{G}_1, \mathcal{G}_2']_\theta \subset \prod_{j=0}^{m-1} H^{-3m(1-\theta)+m_j+1/2,(-3m(1-\theta)+m_j+1/2)/m}(\Sigma) \\ \text{or in the dual of the space containing the eventual} \\ \text{global relations,} \end{cases}$$

from which, together with (11.18), we obtain:

Proposition 11.4. *If $u \in L^2(\Sigma)^m$, the trace $\vec{S}y$ belongs to*

$$\prod_{j=0}^{m-1} H^{m_j+\min m_j-3m+1,(m_j+\min m_j-3m+1)/m}(\Sigma)$$

or to the dual of the space containing the eventual global relations.

Note that we always have

$$m_j + \min m_j - 3m + 1 < 0,$$

with equality in the only case: $m = 1$, $m_0 = 1$ (which is not an exceptional case). Therefore:

Proposition 11.5. *If $m = 1$, $m_0 = 1$, we may take*

$$\text{(11.23)} \qquad K_2 = L^2(\Sigma).$$

For the other cases, we may take

$$\text{(11.24)} \qquad K_2 = \prod_{j=0}^{m-1} H^{-(3m-m_j-\min m_j-1)}(\Sigma)\ ^{((1))}. \quad \square$$

((1)) For positive s, we set $H^{-s}(\Sigma) = (H_{,0}^{s,s}(\Sigma))'$.

11.4 Examples

Example 11.1

The state y is given by

$$(11.25)\qquad \begin{cases} -\Delta y + y'' = 0. \\ \dfrac{\partial y}{\partial \nu} = u, \\ y(x,0) = 0, \quad y'(x,0) = 0. \end{cases}$$

We have: $m = 1$, $m_0 = 1$; according to Propositions 11.1 and 11.5 we may take

$$K_1 = L^2(Q), \quad K_2 = L^2(\Sigma).$$

The *adjoint problem is*:

$$(11.26)\qquad \begin{cases} -\Delta p + p'' = \beta_1(y - y_d^1), \\ \dfrac{\partial p}{\partial \nu} = \beta_2(y|_\Sigma - y_d^2), \\ p(x,T) = 0, \quad p'(x,T) = 0. \quad \square \end{cases}$$

Example 11.2

The state y is given by

$$(11.27)\qquad \begin{cases} -\Delta y + y'' = 0, \\ y|_\Sigma = u, \\ y(x,0) = 0, \quad y'(x,0) = 0. \end{cases}$$

Then $m = 1$, $m_0 = 0$. According to (11.10), we take

$$K_1 = H^{-1}(0,T; H^0(\Omega))$$

and according to (11.24):

$$K_2 = H^{-2}(\Sigma).$$

Let Δ_Γ be the Laplace-Beltrami operator on Γ. Then

$$(11.28)\qquad \begin{cases} \Lambda_2 = (\text{canonical isomorphism of } K_2 \text{ onto } K_2') \\ \quad = \left(-\Delta_\Gamma - \dfrac{\partial^2}{\partial t^2}\right)^{-2}, \\ \text{with Dirichlet boundary conditions,} \end{cases}$$

i.e. if $f \in K_2$, $\Lambda_2 f = \omega$ is the solution of

$$\text{(11.29)} \quad \begin{cases} \left(-\Delta_\Gamma - \dfrac{\partial^2}{\partial t^2}\right)^2 \omega = f, \\ \omega(x, 0) = \omega'(x, 0) = 0, \quad \omega(x, T) = \omega'(x, T) = 0. \end{cases}$$

Then, the *adjoint problem* is

$$\text{(11.30)} \quad \begin{cases} -\Delta p + p'' = \beta_1 \Lambda_1 (y - y_d^1), \quad \Lambda_1 \text{ defined by (11.11)}, \\ p|_\Sigma = \beta \Lambda_2 \left(\dfrac{\partial y}{\partial \nu} - y_d^2\right), \quad \Lambda_2 \text{ defined by (11.29)}, \\ p(x, T) = 0, \quad p'(x, T) = 0. \end{cases}$$

Example 11.3

The state y is given by

$$\text{(11.31)} \quad \begin{cases} \Delta^2 y + y'' = 0, \\ y|_\Sigma = u_1, \quad \left.\dfrac{\partial y}{\partial \nu}\right|_\Sigma = u_2, \quad u = \{u_1, u_2\}, \\ y(x, 0) = 0, \quad y'(x, 0) = 0. \end{cases}$$

Then:

$$m = 2, \quad m_0 = 0, \quad m_1 = 1.$$

We still take K_1 according to (11.10). By (11.24):

$$K_2 = H^{-5}(\Sigma) \times H^{-4}(\Sigma),$$

and the adjoint problem is

$$\text{(11.32)} \quad \begin{cases} \Delta^2 p + p'' = \beta_1 \Lambda_1 (y - y_d^1), \\ \left(-\Delta_\Gamma - \dfrac{\partial^2}{\partial t^2}\right)^5 p|_\Sigma = \beta_2 \left(\left.\dfrac{\partial \Delta y}{\partial \nu}\right|_\Sigma - y_{d_1}^2\right), \\ \qquad p, \ldots, \dfrac{\partial^4 p}{\partial t^2} \text{ zero at } 0 \text{ and } T, \\ \left(-\Delta_\Gamma - \dfrac{\partial^2}{\partial t^2}\right)^4 \left.\dfrac{\partial p}{\partial \nu}\right|_\Sigma = \beta_2 (-\Delta y|_\Sigma - y_{d_2}^2), \\ \dfrac{\partial p}{\partial \nu}, \ldots, \dfrac{\partial^3}{\partial t^3}\left(\dfrac{\partial p}{\partial \nu}\right) \text{ zero at } 0 \text{ and } T, \\ p(x, T) = 0, \quad p'(x, T) = 0. \end{cases}$$

Example 11.4

The state y is given by

$$(11.33)\qquad \begin{cases} \Delta^2 y + y'' = 0, \\ \Delta y\,|_\Sigma = u_1, \quad \left.\dfrac{\partial \Delta y}{\partial \nu}\right|_\Sigma = u_2, \\ y(x,0) = 0, \quad y'(x,0) = 0. \end{cases}$$

Then $m = 2$, $m_0 = 2$, $m_1 = 3$. We still take K_1 according to (11.10) and, according to (11.24), we take

$$K_2 = H^{-1}(\Sigma) \times L^2(\Sigma).$$

And the adjoint problem is given by

$$(11.34)\qquad \begin{cases} \Delta^2 p + p'' = \beta_1 \Lambda_1 (y - y_d^1), \quad \Lambda_1 \text{ defined by (11.11)}, \\ \left(-\Delta_\Gamma - \dfrac{\partial^2}{\partial t^2}\right)(\Delta p\,|_\Sigma) = \beta_2 \left(-\left.\dfrac{\partial y}{\partial \nu}\right|_\Sigma - y^2_{d_1}\right), \\ \Delta p \text{ zero at } 0 \text{ and } T, \\ \dfrac{\partial}{\partial \nu} \Delta p\,|_\Sigma = \beta_2 (y\,|_\Sigma - y^2_{d_2}), \quad \dfrac{\partial \Delta p}{\partial \nu} \text{ zero at } 0 \text{ and } T, \\ p(x,T) = 0, \quad p'(x,T) = 0. \end{cases}$$

12. Comments

The bibliography for control problems for systems governed by *ordinary* differential equations is enormous; the basic reference is Pontryagin-Boltyanskii-Gamkrelidze-Mischenko [1].

Most control problems considered in the literature pertaining to systems governed by ordinary differential equations lead to analogous problem *statements* for systems governed by partial differential evolution equations. The subject is therefore immense, and its analysis is only in its beginnings.

Thus, the present chapter necessarily has a very *incomplete* aspect. We study only *linear* systems, with fixed terminal time and quadratic cost function. For *optimal time* and *controllability* problems, consult Balakrishnan [1], Conti [1], Ju. V. Egorov [1, 2], Falb [1], Fattorini [1–3].

In this chapter, we present the problems resulting from the study of control problems when, in particular, the controls appear in the boundary conditions (boundary control) (the necessity of using the theory of non-homogeneous boundary value problems being evident in this case).

More general results are given in Lions [26]; we have restricted the present discussion to certain points which we believe to be rather significant: the essential result is that, after (rather considerable and seemingly unavoidable) technical difficulties bound to the *justification* of Green's formula used in each case, we are led to new types of boundary value problems (see, for example, Theorem 6.2), which replace the "two-point boundary value problems" of the usual theory for systems governed by ordinary differential equations – see Lions [26, 28, 30].

For problems of *existence* of optimal controls for systems governed by partial differential equations, consult Cecconi [1], Cesari [1, 2], Lions [24, 25], Schmaedeke [1], Vallee [1], Zolezzi (1]. For systems described by elliptic equations considered from the point of view of this chapter, see Lions [26], note (II) and [30].

13. Problems

We only indicate *some* of the very numerous, still unsolved problems in the direction taken in this chapter.

13.1 Extension of the situation of Section 11 to the case of non-zero E_1, E_3, E_4 and where $J(u)$ contains norms in K_3 and K_4.

13.2 The preceding considerations extend, without difficulty, to the Schroedinger equations

$$\mathrm{i}\, A\, y + y' = E_1\, u_1,$$

with the initial and boundary conditions corresponding to Chapter 5, Section 10 (in this case, consider *complex* Hilbert spaces and replace (3.5) with

$$\operatorname{Re} \pi(u^0, u - u^0) \geqq \operatorname{Re} M(u - u^0), \qquad \forall u \in \mathscr{U}).$$

However, the best choice of spaces seems rather delicate.

13.3 In general, we shall obtain an analogous theory to the one in this chapter, *every time* a new class of non-homogeneous evolution problems will be solved.

Consequently, we shall be able to study optimal control problems in connection with problems pertaining to *coupled equations*, *equations with transmission boundary conditions*, etc., as stated in Chapter 4, Section 17 and in Chapter 5, Section 14.

It will also be of interest to study control problems (especially when the control appears in the boundary conditions) in connection with first-order hyperbolic systems, general parabolic equations, coupled systems (first-order hyperbolic, parabolic). For first-order hyperbolic systems, see Russell [1], Schmaedeke [1] and partial results in Lions [30].

Appendix

Boundary Value Problems and Operator Extensions

1. Statement of the Problem. Well-Posed Spaces

1.1 Notation

Let: E be a Banach space on $\mathbf{C}$, $\mathscr{F}$ be a locally convex, topological vector space such that $E \subset \mathscr{F}$ and P be given with:

$$P \in \mathscr{L}(E;\mathscr{F}). \tag{1.1}$$

Set

$$E_P = \{u \mid u \in E, P u \in E\}. \tag{1.2}$$

Provide E_p with the norm

$$\|u\| + \|P u\| = \|u\|_{E_P}, \tag{1.3}$$

where

$$\|\,.\,\| = \text{norm in } E;$$

this makes E_P a *Banach space* (indeed, it is easy to verify that the space E_P is complete in the norm $\|u\|_{E_P}$).

Roughly speaking, the problem to be studied here is the following: *find all closed vector subspaces U of E_P such that*

$$P \text{ is an isomorphism of } U \text{ onto } E. \tag{1.4}$$

Such U's will be called *well-posed spaces* of E_P; if (1.4) holds, the problem (the boundary value problem for the applications we have in mind):

$$P u = f, \quad f \text{ given in } E, \quad u \in U, \tag{1.5}$$

admits a unique solution which depends continuously on f.

Evidently, we cannot expect results different from the definition unless we *assume*:

(1.6) *there exists a closed subspace U_1 of E_P such that P is an isomorphism of U_1 onto E.*

1.2 First Condition for U to be Well-Posed

We introduce

$$Z = \{u \mid u \in E, P u = 0\}. \tag{1.7}$$

Then:

Proposition 1.1. *Assume that* (1.6) *holds.*

A necessary and sufficient condition for U to be a well-posed space is that there exists a k with

$$k \in \mathcal{L}(U_1; Z), \tag{1.8}$$

such that

$$U = (I + k)\, U_1 \tag{1.9}$$

(i. e. U = image of U_1 under $I + k$, I = identity).

Proof. 1) *The condition is necessary.*

Let $u_1 \in U_1$. Assume U to be well-posed. Then there exists a unique $u \in U$, solution of

$$P u = P u_1. \tag{1.10}$$

Therefore

$$\zeta = u - u_1 \in Z$$

and

$$u_1 \to \zeta$$

is a continuous linear mapping of $U_1 \to Z$; let k be this mapping. Then

$$u = u_1 + \zeta = (I + k)\, u_1,$$

and the result follows (since, as u_1 describes U_1, $P u_1$ describes E, therefore $P u$ describes E and finally u describes U).

2) *The condition is sufficient.*

Now let the space U be *defined* by (1.9). We verify that U is *closed* in E_P. Indeed, let $u_n \in U$ with

$$u_n \to u \quad \text{in } E, \qquad P u_n \to f \quad \text{in } E.$$

By definition

$$u_n = v_n + k\, v_n, \qquad v_n \in U_1,$$

and therefore

$$P u_n = P v_n \to f.$$

Since P is an isomorphism of U_1 onto E, we see that

$$v_n \to v \quad \text{in } U_1$$

and then

$$u = v + k\,v \in U \quad \text{and} \quad f = P\,u; \qquad \text{therefore } U \text{ is closed.}$$

It remains to be shown that P is an *isomorphism* of U onto E. Let $e \in E$. We seek $u \in U$ with

$$P\,u = e. \tag{1.11}$$

But, by hypothesis, there exists a $v \in U_1$ with $u = v + k\,v$. Then $P\,u = P\,v$, therefore $P\,v = e$, therefore v exists and is unique, therefore (1.11) admits a *unique solution in* U. □

1.3 The Space D_0

We consider a *closed subspace* D_0 *of* U_1:

$$D_0 \subset U_1 \tag{1.12}$$

and we seek the well-posed spaces U such that *in addition*

$$D_0 \subset U. \tag{1.13}$$

Remark 1.1. In the examples, D_0 will be the closure in E_P of the space of infinitely differentiable functions with compact support in the domain in which the boundary value problems are studied. □

Remark 1.2. Let P_0 be the restriction of P to D_0. The search for well-posed U's such that (1.13) holds is equivalent to the search for all *extensions of* P_0 *which are isomorphisms.* □

We have:

Proposition 1.2. *Assume that* (1.6) *and* (1.12) *are satisfied. A necessary and sufficient condition for* U *to be a well-posed space containing* D_0 *is that* U *be of the form* (1.9), *with* k *satisfying* (1.8) *and furthermore*

$$k = 0 \quad \text{on} \quad D_0. \tag{1.14}$$

Proof. 1) Let U be of the form (1.9) with (1.14). Then, if $d_0 \in D_0$,

$$d_0 + k\,d_0 = d_0 \in U, \quad \text{therefore} \quad D_0 \subset U.$$

2) Conversely, let $D_0 \subset U$. Let us again consider the *construction* of k given in 1) of the proof of Proposition 1.1. Let $d_0 \in D_0$; we consider the solution in U of

$$P\,u = P\,d_0 \tag{1.15}$$

and then

$$k\,d_0 = u - d_0.$$

Since $D_0 \subset U$, *the* solution of (1.15) in U is $u = d_0$, so that $k\,d_0 = 0$. □

Following result (1.14) it is, of course, advisable to *pass to the quotient by* D_0. Thus, we introduce

$$U_1^\bullet = U_1/D_0 \tag{1.16}$$

and the quotient mapping of k:

$$k^\bullet \in \mathscr{L}(U_1^\bullet; Z). \tag{1.17}$$

Then

Corollary 1.1. *Assume that* (1.6) *and* (1.12) *are satisfied. A necessary and sufficient condition for U to be a well-posed space containing* D_0 *is that there exists a* $k^\bullet \in \mathscr{L}(U_1^\bullet; Z)$ *such that*

$$U = U_1 + k^\bullet\, U_1^\bullet. \quad \square \tag{1.18}$$

2. Abstract Boundary Conditions

2.1 Boundary Spaces and Operators

We are given Banach spaces Φ, Ψ and operators β, π, such that

$$\beta \in \mathscr{L}(U_1; \Phi), \quad \beta \text{ surjective with kernel } D_0, \tag{2.1}$$

$$\pi \in \mathscr{L}(E_P; \Psi), \quad \pi \text{ surjective with kernel } U_1. \tag{2.2}$$

The spaces Φ, Ψ are *the boundary spaces* and the operators β, π *the boundary operators*.

Let $\beta^\bullet$ be the quotient operator of β by D_0:

$$\beta^\bullet \in \mathscr{L}(U_1^\bullet; \Phi), \quad \beta^\bullet, \quad \textit{isomorphism of } U_1^\bullet \textit{ onto } \Phi. \tag{2.3}$$

Let π_Z be the restriction of π to Z. Since we have the decomposition into direct sum

$$\begin{cases} E_P = U_1 + Z, \\ e = u_1 + \zeta, \quad P u_1 = P e, \quad u_1 \in U_1, \end{cases} \tag{2.4}$$

we see that

$$\pi_Z \text{ is } \textit{an isomorphism of } Z \textit{ onto } \Psi. \tag{2.5}$$

2.2 Characterization of Well-Posed Spaces

We are now in a position to prove the main "abstract" theorem:

Theorem 2.1. *Assume that* (1.6) *and* (1.12) *are satisfied. Assume that we are given the boundary spaces and operators with* (2.1), (2.2). *A necessary* *and sufficient condition for U to be a well-posed space containing D_0 is that there exists a K with*

$$K \in \mathscr{L}(\Phi, \Psi), \tag{2.6}$$

in such a way that U is the space of u's $\in E_P$ such that

$$\pi u = K \beta (u - \pi_Z^{-1} \pi u). \tag{2.7}$$

Remark 2.1. Condition (2.7) has meaning, since

$$u - \pi_Z^{-1} \pi u \in U_1 \tag{2.8}$$

by virtue of the fact that

$$\pi (u - \pi_Z^{-1} \pi u) = 0. \quad \square$$

Proof of Theorem 2.1. 1) Let U be well-posed and contain D_0. Then, there exist

$$k \in \mathscr{L}(U_1; Z), \quad k^\bullet \in \mathscr{L}(U_1^\bullet; Z), \quad \text{with (1.18)}.$$

Set

$$K = \pi_Z k^\bullet (\beta^\bullet)^{-1}; \tag{2.9}$$

K belongs to $\mathscr{L}(\Phi; \Psi)$.

Then every $u \in U$ may be written:

$$u = u_1 + \zeta, \quad \zeta = k^\bullet u_1^\bullet, \quad u_1 \in U_1. \tag{2.10}$$

Since $u_1 \in U_1$, we have: $\pi u_1 = 0$, therefore

$$\pi u = \pi \zeta \tag{2.11}$$

and

$$\zeta = \pi_Z^{-1} \pi u. \tag{2.12}$$

But, according to (2.9), $k = \pi_Z^{-1} K \beta^\bullet$, therefore

$$\pi \zeta = \pi k^\bullet u_1^\bullet = K \beta^\bullet u_1^\bullet = K \beta u_1 = K \beta (u - \zeta),$$

which, together with (2.11), (2.12), yields (2.7). $\square$

2) Conversely, let U be the (closed) space of elements $u \in E_P$ satisfying (2.7). It follows immediately that $D_0 \subset U$.

Let us this time *define* $k^\bullet$ by

$$k^\bullet = \pi_Z^{-1} K \beta^\bullet, \quad k^\bullet \in \mathscr{L}(U_1^\bullet; Z);$$

and for u given in U, let us define:

$$\zeta = \pi_Z^{-1} \pi u \in Z,$$

$$u_1 = u - \pi_Z^{-1} \pi u = u - \zeta \in U_1 \quad \text{(see Remark 2.1).}$$

Then

$$u - (u_1 + k^{\bullet} u_1^{\bullet}) = \zeta - k^{\bullet} u_1^{\bullet} \in Z. \tag{2.13}$$

We verify that

$$\pi[u - (u_1 + k^{\bullet} u_1^{\bullet})] = 0. \tag{2.14}$$

Indeed

$$\pi[u - (u_1 + k^{\bullet} u_1^{\bullet})] = \pi u - \pi k^{\bullet} u_1^{\bullet}. \tag{2.15}$$

Now, by (2.7):

$$\pi u = K\beta(u - \pi_Z^{-1} \pi u) = K\beta u_1 = K\beta^{\bullet} u_1^{\bullet} = \pi_Z k^{\bullet} u_1^{\bullet} = \pi k^{\bullet} u_1^{\bullet},$$

therefore (2.15) implies (2.14).

But, according to (2.5), it follows from (2.13), (2.14) that

$$u = u_1 + k^{\bullet} u_1^{\bullet}.$$

This completes the proof of the theorem. □

3. Example 1. Elliptic Operators

3.1 Notation

We adopt the notation of Chapter 2.

Let Ω be a bounded open set in $\mathbf{R}^n$, with infinitely differentiable boundary Γ, and let A be a *properly elliptic* operator in $\bar{\Omega}$, with regular coefficients.

In the notation of Sections 1 and 2, we choose:

$$\begin{cases} E = L^2(\Omega), & \mathscr{F} = \mathscr{D}'(\Omega), \\ P = A. \end{cases} \tag{3.1}$$

We take

$$U_1 = H^{2m}(\Omega) \cap H_0^m(\Omega) \tag{3.2}$$

and we assume that

(3.3) A is an isomorphism of U_1 onto $L^2(\Omega)$.

Remark 3.1. Thus we assume the Dirichlet problem to be well-posed for the operator A. We know (see Chapter 2, Section 8), for example if A is strongly elliptic, that $A + \lambda$ is an isomorphism of U_1 onto $L^2(\Omega)$ for sufficiently large real λ. We can therefore apply the present theory to $A + \lambda$. □

The *space* D_0 is chosen as follows:

$$D_0 = \text{closure of } \mathscr{D}(\Omega) \text{ in } E_P = \{u \mid u, A\,u \in L^2(\Omega)\}.$$

We have (see Chapter 2, Section 5):

$$D_0 = H_0^{2m}(\Omega). \tag{3.4}$$

3.2 The Boundary Operators and Spaces

For $u \in U_1$, we set

$$\beta\,u = \left\{\frac{\partial^j u}{\partial \nu^j} \,\middle|\, m \leqq j \leqq 2m-1\right\}, \tag{3.5}$$

$$\Phi = \prod_{j=m}^{2m-1} H^{2m-j-1/2}(\Gamma) \tag{3.6}$$

(see Chapter 1, Section 7 for these spaces).

Condition (2.1) is satisfied by Chapter 1, Section 8.2.

Next, for $u \in E_P$, let

$$\pi\,u = \left\{\frac{\partial^j u}{\partial \nu^j} \,\middle|\, 0 \leqq j \leqq m-1\right\}, \tag{3.7}$$

$$\Psi = \prod_{j=0}^{m-1} H^{-j-1/2}(\Gamma). \tag{3.8}$$

Condition (2.2) is satisfied according to Chapter 2, Section 6. Note that, changing j to $2m - j - 1$,

$$\Phi = \prod_{j=0}^{m-1} H^{j+1/2}(\Gamma),$$

so that

$$\Psi = \Phi'.$$

3.3 Consequences

Let us write out (2.7) for the present situation.

Let u be given in $L^2(\Omega)$, with $A\,u \in L^2(\Omega)$. Let

$$\begin{cases} K = \| K_{jl} \| \\ K_{jl} \in \mathscr{L}(H^{2m-l-1/2}(\Gamma); H^{-j-1/2}(\Gamma)) \end{cases} \tag{3.9}$$

and let ζ be *the* solution of

$$\left\{\begin{aligned} & A\,\zeta = 0 \\ & \frac{\partial^j \zeta}{\partial \nu^j} = \frac{\partial^j u}{\partial \nu^j}, \quad 0 \leqq j \leqq m-1. \end{aligned}\right. \tag{3.10}$$

Then $u - \zeta \in H^{2m}(\Omega)$ and U *is defined as the space of elements* $u \in L^2(\Omega)$ *such that* $A\,u \in L^2(\Omega)$, *which satisfy*

$$\frac{\partial^j u}{\partial \nu^j} = \sum_{l=m}^{2m-1} K_{jl} \frac{\partial^l}{\partial \nu^l}(u - \zeta), \quad 0 \leqq j \leqq m-1. \tag{3.11}$$

By varying K in (3.9), we obtain *all* well-posed spaces.

3.4 Various Remarks

Remark 3.2. In (3.1) we started from hypothesis (3.3).

We could assume that *another* problem, different from the Dirichlet problem, is well-posed. For example (in the notation of Chapter 2), assume that the problem

$$\left\{\begin{aligned} & A\,u = f, \quad f \text{ given in } L^2(\Omega), \quad u \in H^{2m}(\Omega), \\ & B_j\,u = 0, \quad 0 \leqq j \leqq m-1, \end{aligned}\right. \tag{3.12}$$

is *well-posed.*

In that case, we take

$$U_1 = \{u \mid u \in H^{2m}(\Omega), B_j\,u = 0, 0 \leqq j \leqq m-1\}, \tag{3.13}$$

$$\beta u = \{S_j u \mid 0 \leqq j \leqq m-1\} \text{ (notation of Chapter 2, Section 2).} \tag{3.14}$$

If the order of $S_j = \mu_j$, we have

$$\Phi = \prod_{j=0}^{m-1} H^{2m-\mu_j-1/2}(\Gamma). \tag{3.15}$$

Next, for $u \in E_P$, we take

$$\pi\,u = \{B_j\,u \mid 0 \leqq j \leqq m-1\}, \tag{3.16}$$

$$\Psi = \prod_{j=0}^{m-1} H^{-m_j-1/2}(\Gamma). \tag{3.17}$$

All well-posed spaces are given by the elements $u \in L^2(\Omega)$ such that $A\,u \in L^2(\Omega)$ and

$$B_j\,u = \sum_{l=0}^{m-1} K_{jl}\,S_l(u - \zeta), \quad 0 \leqq j \leqq m-1, \tag{3.18}$$

where

$$K_{jl} \in \mathscr{L}(H^{2m-\mu_l-1/2}(\Gamma); H^{-m_j-1/2}(\Gamma)), \tag{3.19}$$

$$\zeta \text{ given by } A\zeta = 0, \quad B_j\zeta = B_j u, \quad 0 \leqq j \leqq m-1. \quad \square \tag{3.20}$$

Remark 3.3. If we assume that problem (3.12) is well-posed for two distinct families $B_j^{(1)}$ and $B_j^{(2)}$ of boundary operators, we can define all well-posed spaces by two families of relations of the type (3.18), with two families $K_{jl}^{(1)}$ and $K_{jl}^{(2)}$ of operators. Since the well-posed spaces are evidently the same starting from the two families, there is a one-to-one correspondence between $K^{(1)}$ and $K^{(2)}$ yielding the same space. □

Remark 3.4. We started *from the choice* $E = L^2(\Omega)$. We could also start *from the (Hilbert) choice:*

$$E = H^s(\Omega), \quad s > 0. \tag{3.21}$$

In that case, we consider problem (3.12) with $f \in H^s(\Omega)$. We assume this problem to be well-posed. Then

$$U_1 = \{u \mid u \in H^{s+2m}(\Omega), B_j u = 0, 0 \leqq j \leqq m-1\} \tag{3.22}$$

and this time (compare with (3.15))

$$\Phi = \Phi^s = \prod_{j=0}^{m-1} H^{2m+s-\mu_j-1/2}(\Gamma) \tag{3.23}$$

and β is given by (3.14). Next, with π given by (3.16),

$$\Psi = \Psi^s = \prod_{j=0}^{m-1} H^{s-m_j-1/2}(\Gamma). \tag{3.24}$$

All well-posed spaces are given by (3.18), *where* ζ *is still given by* (3.20), *but instead of* (3.19), *we have*

$$K_{jl} \in \mathscr{L}(H^{2m+s-\mu_l-1/2}(\Gamma); H^{s-m_j-1/2}(\Gamma)). \quad \square \tag{3.25}$$

Remark 3.5. If we compare Remarks 3.2 and 3.4, then we see that the well-posed spaces *depend on the initial choice* of E. For example, for K_{jl}, we can take a *multiplier* of

$$H^{2m-\mu_l-1/2}(\Gamma) \to H^{-m_j-1/2}(\Gamma)$$

which *is not* a multiplier of

$$H^{2m+s-\mu_l-1/2}(\Gamma) \to H^{s-m_j-1/2}(\Gamma). \quad \square$$

Remark 3.6. The boundary conditions (3.11) or (3.18) *are not in general of a local nature.* □

Remark 3.7 (This remark uses notions which are not fully developed in this book).

We can also choose

$$E = L^p(\Omega), \quad 1 < p < \infty, \quad p \neq 2,$$

and assume that (3.12) is well-posed in $L^p(\Omega)$. Then

$$u \in W^{2m,p}(\Omega) = \{v \mid D^\alpha v \in L^p(\Omega), |\alpha| \leqq 2m\}$$

and

$$U_1 = \{u \mid u \in W^{2m,p}(\Omega); B_j u = 0, 0 \leqq j \leqq m-1\}.$$

The formulas are *the same* as in Remark 3.2, but this time with:

$$\Phi = \prod_{j=0}^{m-1} W^{2m-\mu_j-1/p,p}(\Gamma)$$

$$\Psi = \prod_{j=0}^{m-1} W^{-m_j-1/p,p}(\Gamma)$$

(for the spaces $W^{s,p}(\Gamma)$, see for example Lions-Magenes [1]).

The well-posed boundary value problems *depend on* p. □

4. Example 2. Parabolic Operators

4.1 Notation

Now, consider the setting of Chapter 4.

Let Q be the cylinder

(4.1) $$Q = \Omega \times]0, T[$$

and P the operator

(4.2) $$P = A\left(x, t, \frac{\partial}{\partial x}\right) + \frac{\partial}{\partial t} = A + \frac{\partial}{\partial t},$$

with the hypotheses of Chapter 4, Section 1.

We *choose* (notation of Sections 1 and 2)

(4.3) $$E = L^2(Q), \quad \mathscr{F} = \mathscr{D}'(Q).$$

Then

(4.4) $$E_P = \left\{u \mid u \in L^2(Q), A u + \frac{\partial u}{\partial t} \in L^2(Q)\right\}.$$

We take

$$(4.5)\qquad U_1 = \left\{u \mid u \in H^{2m,1}(Q),\, u(x,0) = 0, \frac{\partial^j u}{\partial \nu^j} = \right.$$
$$\left. = 0 \text{ on } \Sigma, 0 \leqq j \leqq m-1\right\}$$

(Recall that Σ is the lateral boundary of Q:

$$\Sigma = \Gamma \times]0, T[).$$

The operator P being assumed parabolic, we know (Chapter 4, Section 1) that

$$(4.6)\qquad P \text{ is an isomorphism of } U_1 \text{ onto } L^2(Q).$$

The space D_0 is chosen as

$$D_0 = \text{closure of } \mathscr{D}(Q) \text{ in } E_P.$$

We have:

Lemma 4.1. *The space D_0 coincides with the closure of $\mathscr{D}(Q)$ in $H^{2m,1}(Q)$, that is*

$$(4.7)\qquad \begin{cases} D_0 = H^{2m,1}_{0,0}(Q) = \left\{v \mid v \in H^{2m,1}(Q), \dfrac{\partial^j v}{\partial \nu^j} = 0 \text{ on } \Sigma,\right. \\ \left. \text{for } 0 \leqq j \leqq 2m-1,\, v(x,0) = 0,\, v(x,T) = 0\right\}. \end{cases}$$

Proof. Indeed (see Chapter 4, Section 1), we have

$$\| P\varphi \|_{L^2(Q)} \geqq c \|\varphi\|_{H^{2m,1}(Q)} \quad \forall \varphi \in \mathscr{D}(Q). \quad \square$$

4.2 The Boundary Operators and Spaces

For $u \in U_1$, we set:

$$(4.8)\qquad \beta u = \left\{\frac{\partial^j u}{\partial \nu^j} \text{ on } \Sigma, m \leqq j \leqq 2m-1; u(x,T)\right\}.$$

And, we let

$$(4.9)\qquad \Phi = \left\{\prod_{j=m}^{2m-1} H^{2m-j-1/2,(2m-j-1/2)/2m}(\Sigma) \times H_0^m(\Omega)\right\}.$$

Then *condition* (2.1) *is satisfied.*

For $u \in E_P$, we define (formally at first) πu by

(4.10) $$\pi u = \left\{\frac{\partial^j u}{\partial \nu^j} \text{ on } \Sigma, 0 \leqq j \leqq m-1; u(x,0)\right\}.$$

We know (Chapter 4, Section 10) that for $u \in E_P$, we have

(4.11) $$\begin{cases} \left.\dfrac{\partial^j u}{\partial \nu^j}\right|_\Sigma \in H^{-(j+1/2),-(j+1/2)/2m}(\Sigma), \quad 0 \leqq j \leqq m-1, \\ u(x,0) \in H^{-m}(\Omega). \end{cases}$$

But we can state this more precisely (see Chapter 4, Section 12.3). *Define:*

(4.12) $$\begin{cases} \Psi_1 = \text{space described by } T_j v,\ 0 \leqq j \leqq m-1, \text{ and} \\ \qquad v(x,0) \text{ as } v \text{ describes } H^{2m,1}(Q), \text{ with} \\ \qquad \dfrac{\partial^j v}{\partial \nu^j} = 0 \text{ on } \Sigma, 0 \leqq j \leqq m-1, \\ \qquad v(x,T) = 0, \end{cases}$$

where T_j corresponds to $B_j = \dfrac{\partial^j}{\partial \nu^j}$ in Green's formula (see Chapter 2, Section 2 and Chapter 4, Section 1).

Therefore

(4.13) $$\Psi_1 = \{\psi_j, \varphi \mid \psi_j \in H^{j+1/2,(j+1/2)/2m}(\Sigma), \varphi \in H_0^m(\Omega)\}.$$

Then we can define

(4.14) $$\pi u = \left\{\frac{\partial^j u}{\partial \nu^j}, u(x,0)\right\} \in \Psi_1'$$
$$= \prod_{j=0}^{m-1} H^{-j-1/2,-(j+1/2)/2m}(\Sigma) \times H^{-m}(\Omega)$$

with the help of *Green's formula* and of a continuous linear right-inverse of

$$\Psi_1 \to \left\{v \mid v \in H^{2m,1}(Q), \frac{\partial^j v}{\partial \nu^j} = 0 \text{ on } \Sigma, 0 \leqq j \leqq m-1, v(x,T) = 0\right\}.$$

Now, we can take

(4.15) $$\Psi = \Psi_1',$$

and condition (2.2) *is satisfied.*

4.3 Consequences

Let u be given in E_P. We define ζ by

$$(4.16)\qquad \begin{cases} A\zeta + \dfrac{\partial \zeta}{\partial t} = 0, \\[2ex] \dfrac{\partial^j \zeta}{\partial \nu^j} = \dfrac{\partial^j u}{\partial \nu^j}, \quad 0 \leqq j \leqq m-1, \\[2ex] \zeta(x, 0) = u(x, 0), \end{cases}$$

(not formally, but in the sense of Remark 12.3 of Chapter 4).

Then, if $K \in \mathscr{L}(\Phi; \Psi)$, the well-posed space U corresponding to K is defined by

$$(4.17)\qquad \pi u = K\beta(u - \zeta),$$

which decomposes into

$$(4.18)\qquad \begin{cases} \dfrac{\partial^j u}{\partial \nu^j} = \displaystyle\sum_{l=m}^{2m-1} K_{jl} \dfrac{\partial^l (u - \zeta)}{\partial \nu^l} + K_j(u(x, T) - \zeta(x, T)), \\[2ex] 0 \leqq j \leqq m-1, \ u(x, 0) = \displaystyle\sum_{l=m}^{2m-1} \tilde{K}_l \dfrac{\partial^l (u - \zeta)}{\partial \nu^l} + \\[2ex] + \tilde{K}(u(x, T) - \zeta(x, T)). \end{cases}$$

In this way, we obtain *all* well-posed spaces. □

5. Example 3. Evolution Operators of the Second Order in t

5.1 Notation

We now consider the setting of Chapter 5. The cylinder Q is defined as in Section 4 and

$$(5.1)\qquad P = A\left(x, t, \frac{\partial}{\partial x}\right) + \frac{\partial^2}{\partial t^2},$$

with the hypotheses of Chapter 5.

The interpretation of the abstract results are delicate, because of the difficulties with the regularity and the trace theorems noted in Chapter 5.

5.2 Formal Results

Formally, we take

(5.2)

$$U_1 = \left\{ u \mid u \in H^{m,1}(Q),\, P\,u \in L^2(Q),\, \frac{\partial^j u}{\partial \nu^j} = 0,\, 0 \leqq j \leqq m-1, \right.$$

$$\left. u(x,0) = 0,\, \frac{\partial u}{\partial t}(x,0) = 0 \right\},$$

(5.3) $$\beta\,u = \left\{ \frac{\partial^j u}{\partial \nu^j} \text{ on } \Sigma,\, m \leqq j \leqq 2m-1,\, u(x,T),\, \frac{\partial u}{\partial t}(x,T) \right\},$$

(5.4) $$\pi\,u = \left\{ \frac{\partial^j u}{\partial \nu^j} \text{ on } \Sigma,\, 0 \leqq j \leqq m-1,\, u(x,0),\, \frac{\partial u}{\partial t}(x,0) \right\}.$$

The space Φ is the "trace" space described by $\beta\,u$ as u describes U_1, and the space Ψ is the "trace" space described by $\pi\,u$ as u describes

$$E_P = \{u \mid u \in L^2(Q),\, P\,u \in L^2(Q)\}.$$

For u given in E_P, the function ζ is an (appropriately interpreted; see Chapter 5) solution of

(5.5) $$\begin{cases} A\,\zeta + \dfrac{\partial^2 \zeta}{\partial t^2} = 0, \\[2mm] \dfrac{\partial^j \zeta}{\partial \nu^j} = \dfrac{\partial^j u}{\partial \nu^j}, \quad 0 \leqq j \leqq m-1, \quad \text{on } \Sigma, \\[2mm] \zeta(x,0) = u(x,0), \\[2mm] \dfrac{\partial \zeta}{\partial t}(x,0) = \dfrac{\partial u}{\partial t}(x,0). \end{cases}$$

The well-posed spaces are defined by a relation of the form (4.17), where $K \in \mathscr{L}(\Phi; \Psi)$. □

Remark 5.1. Similar (also *formal*) results can be developed for Schroedinger-type operators (see Chapter 5.) □

6. Comments and Problems

"All" extensions leading to well-posed boundary value problems have been specified, for second-order elliptic operator by Vishik [2]. For elliptic operators of order $2m$, $m > 1$, the corresponding problem has been solved by Peetre and Lions (see Peetre [1]).

For points of view different from this Appendix, see Bade-Freeman [1], Beals [1], Birman [1], Browder [7], Calkin [1], Fishel [1], Freeman [2], Friedrichs [4], M. G. Krein [1], Schechter [1]. We also recall the non-local elliptic problems which come up in probability theory and have been noted in the Comments to Chapter 2.

We have given a very general (and very simple) abstract presentation which *also applies to evolution operators*; the results of Section 4 are apparently new. The same is true for Section 5, but the spaces appearing there should be *specified*; this should be possible at least for certain particular cases, but we have not attempted to do so. Similarly, it would be interesting to apply the preceding considerations to *quasi-elliptic* operators.

For *elliptic* operators, a much more complete study is done in Grubb [1]; it would be of interest to extend the results of this author to the *parabolic* case.

Along the same lines, we call attention to the work of Phillips [2] on maximal extensions of contraction operators, and to the work of Carroll [1] which constitutes a kind of "abstract analogue" to Section 4.

Bibliography[1]

Agmon, S.

1. Multiple layer potentials and Dirichlet problem for higher order elliptic equations in the plane. Comm. Pure Appl. Math. **10**, 179—239 (1957).
2. The coerciveness problem for integro-differential forms. J. Analyse Math. **6**, 183—223 (1958).
3. The L_p approach to the Dirichlet problem-I. Ann. Sc. Norm. Sup. Pisa **13**, 405—448 (1959).
4. Remarks on self-adjoint and semi-bounded elliptic boundary value problems. Proc. Intern. Symp. on linear spaces, 1—13, Jerusalem: Pergamon Press 1960.
5. Problèmes mixtes pour les équations hyperboliques d'ordre supérieur, Colloques sur les équations aux dérivées partielles, C.N.R.S., Paris 1962, 13—18.
6. On the eigenfunctions and on the eigenvalues of general elliptic boundary value problems. Comm. Pure Appl. Math. **15**, 119—147 (1962).
7. Lectures on Elliptic Boundary Value problems, Van Nostrand Mathematical Studies, Princeton: 1965.
8. On Kernels, eigenvalues and eigenfunctions of operators related to elliptic problems. Comm. Pure Appl. Math. **18**, 627—663 (1965).
9. Maximum properties for solutions of higher order elliptic equations. Bull. Amer. Math. Soc. **60**, 77—80 (1960).

Agmon, S., Nirenberg, L.

1. Properties of solutions of ordinary differential equations in Banach spaces, Comm. Pure Appl. Math. Surv., **16**, 121—239, (1963).
2. Lower bounds and uniqueness theorems for solutions of differential equations in a Hilbert space. Comm. Pure Applied Maths. **20**, 207—229, (1967).

Agranovich, M. S.

1. Opérateurs elliptiques singuliers intégro-différentiels. Uspehi Mat. Nauk **20**, 3—120 (1965) [Russian Math. Surv. **20**, 1—121 (1965)].
2. Problèmes positifs de type mixte pour certains systèmes hyperboliques. Dokl. Ak. Nauk **167**, 539—542 (1966) [Soviet Math. **7**, 539—542 (1966)].
3. Sur la théorie des problèmes aux limites pour les systèmes symétriques du premier ordre. Mat. Sbornik **73**, 161—197 (1967).

Agranovich, M. S., Vishik M. I.

1. Problèmes elliptiques avec paramètre et problèmes paraboliques de type général. Uspehi Mat. Nauk **19**, No. 3, 53—161 (1964) [Russian Math. Sur., **19**, No. 3, 53—157 (1964)].

[1] Certain references, explicitly used only in Volume I, have been preserved for the convenience of the reader; in particular the references pertaining to evolution equations have been reproduced here.

Albertoni, S., Cerciganni, C.

1. Sur un problème mixte dans la dynamique des fluides. C. R. Ac. Sci. Paris **261**, 312–315 (1965).

Amerio, L., Prouse, G.

1. Abstract almost periodic functions and functional analysis, New York: Van Nostrand 1970.

Arima, R.

1. On general boundary value problems for parabolic equations. Journal of Mathematics of Kyoto University **4**, 207–244 (1964).

Arnese, G.

1. Maggiorazioni in L^p dei potenziali relativi all'equazione del calore. Ric. di Mat. **13**, 1–45 (1964).
2. Su alcune proprietà dei potenziali relativi ad una equazione parabolica d'ordine $2m$. Le Matematiche **20**, 121–135 (1965).

Aronson, D. G.

1. On the Green's function for second order parabolic differential equations with discontinuous coefficients. Bull. Amer. Math. Soc. **69**, 841–847 (1963).
2. Removable singularities for linear parabolic equations. Arch. for Rat. Mech. and Anal. **17**, 78–84 (1964).
3. Isolated singularities of solutions of second order parabolic equations. Arch. for Rat. Mech. and Anal. **19**, 231–238 (1965).

Aronson, D. G., Serrin, J.

1. Local Behavior of Solutions of Quasilinear Parabolic Equations. Archive Rat. Mech. Analysis **25**, 81–122 (1967).
2. A Maximum principle for non linear parabolic equations. Annali Scuola Norm. Sup. Pisa XXI, 291–305 (1967).

Aronszajn, N., Milgram, A. N.

1. Differential operators on Riemannian manifolds. Rend. Circ. Mat. Palermo **2**, 1–61 (1952).

Artola, M.

1. Equations paraboliques à retardement. C. R. Acad. Sc. Paris **264**, A, 668 to 671 (1967).

Bade, W. G., Freeman, R. S.

1. Closed extensions of the Laplace operator determined by a general class of boundary conditions. Pacific J. of Math. **12**, 395–410 (1962).

Baiocchi, C.

1. Su alcuni spazi di distribuzioni e sul problema di Dirichlet per le equazioni lineari ellittiche. Richerche di Mat. **13**, 3–29 (1964).
2. Sui problemi ai limiti per le equazioni paraboliche del tipo del calore. Boll. U.M.I. **3**, 19, 407–422 (1964).
3. Regolarità e unicità della soluzione di una equazione differenziale astratta. Rend. Sem. Mat. Padova **35**, 380–417 (1965).

4. Sul problema misto per l'equazione parabolica del tipo del calore. Rend. Sem. Mat. Padova **36**, 80—121 (1966).
5. Un teorema di interpolazione; applicazioni ai problemi ai limiti per le equazioni a derivate parziali. Ann. Mat. Pura Appl. **4**, LXXIII, 235—252 (1966).
6. Soluzioni ordinarie e generalizzate del problema di Cauchy per equazioni differenziali astratte lineari del primo ordine negli spazi di Hilbert. Ricerche di Mat. **16**, 27—95 (1967).
7. Sulle equazioni differenziali astratte lineari del primo e secondo ordine negli spazi di Hilbert. Ann. Mat. pura appl. **4**, LXXVI, 233—304 (1967).
8. Définition d'operateurs maximaux et applications. Ann. Sc. Ecole Norm. Sup. **4**, 2, 481—520 (1969).

Balakrishnan, A. V.

1. Optimal control problems in Banach spaces. J.S.I.A.M. Control, ser. A, **3**, 152—180 (1965).

Baouendi, M. S.

1. Sur une classe d'opérateurs elliptiques dégénérés, Thèse Paris, 1966. Bull. Soc. Math. France **95**, 45—87 (1967).

Baouendi, M. S., Grisvard, P.

1. Sur une équation d'évolution changeant de type. C. R. Acad. Sc. Paris (1967) [detailed proofs in J. Funct. Anal. **2**, 352—367 (1968)].

Bardos, C.

1. Approximation semi-discrète de la solution d'une équation variationnelle, astreinte à vérifier des conditions aux limites dépendant du temps. Rend. Sem. Mat. Padova **38**, 41—59 (1967).

Barozzi, C.

1. Su una generalizzazione degli spazi $L^{q,\lambda}$ di Morrey, Ann. Sc. Norm. Sup. Pisa **19**, 609—626 (1965).
2. Su certi teoremi di unicità e teoremi di Liouville relativi alle soluzioni di una classe di equazioni ipoellittiche. Atti Semin. Mat. e Fis. Univ. Modena **13**, 41—61 (1964).
3. Un teorema del tipo di Liouville per una classe di equazioni paraboliche. Atti Semin. Mat. e Fis. Univ. Modena **16**, 100—120 (1967).

Beals, R.

1. Non local boundary value problems for elliptic operators. Amer. J. Math. **87**, 315—348 (1965).
2. On eigenvalue distributions for elliptic operators without smooth coefficients. Bull. Amer. Math. Soc. **72**, 701—705 (1966).

Bensoussan, A.

1. L'identification et le filtrage. I.R.I.A., Paris cahier No. 1, 1969.

Birman, M. S.

1. Sur la théorie des prolongements autoadjoints des opérateurs définis positifs. Mat. Sbornik **38**, 431—450 (1956).

Boltyanskii, V., Gamkrelidze, R., Mischenko, L., Pontryagin, L. S.
Cf. Pontryagin-Boltyanskii-Gamkrelidze-Mischenko.

Bourbaki, N.

1. Intégration, chapitre I—IV, Act. Sc. Ind., Paris: Hermann 1965.

Brézis, H.

1. Inéquations d'évolution abstraites. C. R. Acad. Sc. Paris **264**, 732—735 (1967).
2. Equations et inéquations non linéaires dans les espaces vectoriels en dualité. Ann. Inst. Fourier **18**, 115—175 (1968).

Brézis, H., Lions, J. L.

1. Sur certains problèmes unilatéraux hyperboliques. C. R. Acad. Sc. Paris **264**, 928—931 (1967).

Browder, F. E.

1. Estimates and existence theorems for elliptic boundary value problems. Proc. Nat. Acad. Sc. U.S.A. **45**, 365—372 (1959).
2. A priori estimates for solutions of elliptic boundary value problems. Indagationes Math. **22**, 145—169 (1960).
3. Functional Analysis and Partial Differential Equations-I, II. Math. Ann. **138**, 55—79 (1959); **145**, 81—226 (1962).
4. A priori estimates of elliptic and parabolic equations. Proc. Symp. pure math., vol. IV, Partial diff. equat., Amer. Math. Soc. (1961) 73—83.
5. On the spectral theory of elliptic differential operators, I. Math. Ann. **142**, 22—130 (1961).
6. A continuity property for adjoint of closed operators in Banach spaces and its applications to elliptic boundary value problems. Duke Math. J. **28**, 157—182 (1961).
7. Non local elliptic boundary value problems. Amer. J. Math. **86**, 735—750 (1964).
8. Asymptotic distributions of eigenvalues and eigenfunctions for non local elliptic boundary value problems, I, Amer. J. Math. **87**, 175—195 (1965).
9. Strongly elliptic systems of differential equations. Contribution to the Theory of Partial Diff. Equat. Ann. Math. Studies, Princeton (1954), 15—51.
10. Linear parabolic equations. Proc. Nat. Ac. Sci. **49**, 185—190 (1963).
11. Parabolic systems of differential equations with time dependent coefficients. Proc. Nat. Ac. Sci. **42**, 914—917 (1956).

Butzer, P. L.

1. On Dirichlet's problem for the half-space and the behavior of its solution on the boundary. J. Math. Anal. Appl. **2**, 86—96 (1961).

Calderon, A. P.

1. Lebesgue spaces of differentiable functions and distributions, Partial Diff. Equations. Proc. Symp. Pure Math., vol. 4, A.M.S. (1961), 33—49.
2. Intermediate spaces and interpolation, Studia Math., Série spéciale t. I. Coll. Anal. Fonct. de Varsovie (1963), 31—34.
3. Intermediate spaces and interpolation, the complex method. Studia Math. **24**, 113—190 (1964).

4. Boundary value problems for elliptic equations. Proc. Joint Soviet-American Symp. on Part. Diff. Equat., Novosibirsk (1963), 1—4.
5. Algebras of singular integral operators. Lecture delivered at the International Congress, Moscow, 1966, Proceedings, 393—395.
6. Integrales singulares y sus aplicaciones a ecuaciones differenciales hiperbolicas, Universidad de Buenos Aires, fasc. 3, 1960.
7. Uniqueness in the Cauchy problem for partial differential equations. Amer. J. Math. **80**, 16—36 (1958).
8. The analytic calculation of the index of elliptic equations. Proc. Nat. Ac. Sci. **57**, 1193—1194 (1967).

Calkin, J. W.

1. Abstract symmetric boundary conditions. Trans. Amer. Math. Soc. **45**, 369—442 (1939).

Campanato, S.

1. Sui problemi al contorno per sistemi di equazioni differenziali del tipo de l'elasticità, I, II. Ann. Sc. Norm. Sup. Pisa **12**, 223—258, 275—302 (1959).
2. Sulla regolarità delle soluzioni di equazioni differenziali di tipo ellittico, Pisa: Ed. Tecnico Scientifiche **1963**.
3. Proprietà di una famiglia di spazi funzionali. Ann. Sc. Norm. Sup. Pisa **18**, 137—160 (1964).
4. Equazioni ellittiche del secondo ordine e spazi $\mathscr{L}^{2,\lambda}$. Ann. Mat. Pura Appl. **4**, LXIX, 321—382 (1965).
5. Equazioni paraboliche del secondo ordine e spazi $\mathscr{L}^{2,\lambda}(\Omega, \delta)$. Ann. Mat. Pura Appl. **4**, LXXIII, 55—102 (1966).
6. Alcuni osservazioni relative alle soluzioni di equazioni ellittiche d'ordine $2m$. Atti, Conv. Equaz. a deriv. parz., Bologna, aprile 1967, Roma: Cremonese, 1968, 17—25.

Campbell, L.

1. Solution of a mixed problem for a hyperbolic differential equation by Riemann's method. Acta Math. **100**, 23—43 (1958).

Campbell, L., Robinson, A.

1. Mixed problem for hyperbolic partial differential equations. Proc. London Math. Soc. **5**, 129—147 (1945).

Carroll, R.

1. Problems in linked operators, I—II. Math. Annalen **151**, 272—282 (1963); **160**, 233—256 (1965).
2. On the propagator equation. Illinois J. Math., **11**, 506—522 (1967).

Cattabriga, L.

1. Su un problema al contorno relativo al sistema di equazioni di Stokes. Rend. Sem. Mat. Padova **31**, 1—33 (1961).
2. Su un certo spazio funzionale. Un teorema di tracce. Ann. Sc. Norm. Sup. Pisa **20**, 783—796 (1966).
3. Sulla connessione di un teorema di tracce con un certo poliedro convesso. Boll. U.M.I. (3) **22**, 1—12 (1967).
4. Alcuni teoremi di immersione per spazi funzionali generalizzanti gli spazi di S. L. Sobolev. Ann. Univ. Ferrara, Sc. Mat. **12**, 73—88 (1967).

5. Un problema al contorno per una equazione parabolica di ordine dispari. Ann. Sc. Norm. Sup. Pisa **13**, 163—203 (1959).

Cavallucci, A.

1. Sulle proprietà differenziali delle soluzioni delle equazioni quasi-ellittiche. Ann. Mat. Pura Appl. **4**, LXVII, 143—168 (1965).
2. Alcuni teoremi di tracce. Sem. Mat. Fis. di Modena **15**, 137—157 (1966).
3. Costruzione di un rilevamento per la traccia iperpiana dello spazio $H_\mu(\mathbf{R}^n)$. Boll. U.M.I. **22**, 491—496 (1967).
4. Sulla regularita delle soluzione delle equazione quasi-ellittiche in un semispazio. Sem. Mat. Fis. Modena **17**, 1—18 (1967).

Cecconi, J.

1. Sulla teoria dei controlli, Atti Convegno di Analisi funzionale. Ist. Naz. Alta Mat., Roma, Symposia Math., vol. II, 1969, 255—263.

Cercignani, C., Albertoni, S.
Cf. Albertoni-Cercignani.

Cesari, L.

1. Existence theorems for Lagrange and Pontryagin problems of the calculus of variations and optimal control; more dimensional extension in Sobolev space, C.I.M.E., Bressanone **1966**, Calculus of Variations, Classical and Modern, Roma: Cremonese **1967**, 85—176.
2. Existence theorems for multi-dimensional problems of optimal control. Differential Equations and dynamical systems, New York: Academic Press, 1967, 115—132.

Cherruault, Y.

1. Approximation d'opérateurs linéaires et applications, Paris: Dunod **1967**.

Conti, R.

1. On some aspects of linear control theory, Mathem. Theory of Control, edited by Balakrishnan and Neustadt, Acad. Press, 1967, 285—300.

Cooper, J.

1. Un problème relatif aux équations d'évolution abstraites. C. R. Acad. Sc. Paris **265**, 422—424 (1967).

Daletskii, Yu. L.

1. Intégrales continues et équations opérationnelles d'évolution. Uspehi Mat. Nauk **17**, 3—115 (1962) [Russian Math. Surv. **17**, 1—105 (1962)].

Da-Prato, G.

1. Semi-gruppi di crescenza n. Ann. Sc. Norm. Sup. Pisa **20**, 753—782 (1966).
2. Semi gruppi regolarizzabili. Ricerche di Matem. **15**, 223—248 (1966).
3. Spazi $L^{p,\theta}(\Omega, \delta)$ e loro proprietà. Ann. Mat. Pura Appl. **4**, LXIX, 382—395 (1965).
4. Somma di generatori infinitesimali di semigruppi. Ann. Mat. Pura Appl. **4**, 78, 131—158 (1968).

5. Equations opérationnelles dans les espaces de Banach (cas analytique). C. R. Acad. Sc. Paris **266** A, 277—279 (1968).
6. Equazioni differenziali astratte, Pisa: Editrice Tecnico Scientifica 1968.

Da-Prato, G., Giusti, E.

1. Equazioni di evoluzione in L^p. Annali Sc. Norm. Sup. Pisa **21**, 485—505 (1967).
2. Equazioni di Schroedinger e delle onde per l'operatore di Laplace iterato in $L^p(\mathbf{R}^n)$. Annali di Mat. Pura Appl. **4**, LXXXVI, 377—398 (1967).

Da-Prato, G., Mosco, U.

1. Semi-gruppi distribuzioni analitici. Ann. Sc. Norm. Sup. Pisa **19**, 367—396 and 563—576 (1965).

Deprima, C. A.

1. Uniqueness theory of linear hyperbolic partial differential equations, Thesis, New York University 1948.

Dezin, A. A.

1. Théorèmes d'existence et d'unicité pour la solution des problèmes aux limites pour les équations aux dérivées partielles dans des espaces fonctionnels. Uspehi Mat. Nauk **14**, 21—73 (1959) [Amer. Math. Soc. Transl. (2) **42**, 71—128 (1960)].

Diaz, J. B.

1. On singular and regular Cauchy problems. Comm. Pure Appl. Math. **9**, 383—390 (1956).

Doetsch, G.

1. Theorie und Anwendung der Laplace-Transformation, Berlin: Springer 1937.

Dressel, F.

1. The fundamental solutions of the parabolic equations. Duke Math. J. (I) **7**, 186—203 (1940); (II) **13**, 61—70 (1946).

Duff, G. F. D.

1. Mixed problems for hyperbolic equations of general order. Canad. J. Math. **11**, 195—221 (1959).

Dunford, N., Schwartz, J.

1. Linear Operators, I, II, New York: Interscience Publishers 1958, 1963.

Egorov, Ju. V.

1. Conditions nécessaires pour le contrôle optimal dans les espaces de Banach. Mat. Sbornik **64**, 79—101 (1964) [Amer. Math. Soc. Transl. (2) **49**, 63—85 (1966)].
2. Sur certains problèmes de contrôle optimal. J. de Math. numérique et de Physique Math. **3**, 887—904 (1963).

Eidelman, S. D.

1. On fundamental solutions of parabolic systems. Mat. Sbornik **38**, 51—92 (1956) [Amer. Math. Soc. Transl. (2) **41**, 1—48 (1964)].
2. Systèmes paraboliques, Moscow 1964; English translation: Amsterdam: North-Holland 1969.

Eidelman, S. D., Lipko, B.
Cf. Lipko-Eidelman.

Eskin, I. G., Vishik, I. M.
Cf. Vishik-Eskin.

Fabes, E. B.

1. Singular integrals and partial differential equations of parabolic type. Studia Math. **28**, 81—131 (1967).

Fabes, E. B., Jodeit Jr., M.

1. Boundary value problems for second order parabolic equations. Proc. Symp. on singular integrals, Chicago, April 1966; Amer. Math. Soc., vol. **10** (1967).

Fabes, E. B., Rivière, N. M.

1. Systems of parabolic equations with uniformly continuous coefficients. J. Analyse Math. **17**, 306—335 (1966).
2. Singulars integral with mixed homogeneity. Studia Math. **27**, 19—38 (1966).

Faedo, S.

1. Un nuovo metodo per l'analisi esistenziale e quantitativa dei problemi di propagazione. Ann. Sc. Norm. Sup. Pisa **1**, 140, 1—42 (1949).

Falb, P. L.

1. Infinite dimensional control problems I. J. Math. Anal. Appl. **9**, 12—22 (1964).

Fattorini, H. O.

1. Ordinary differential equations in linear topological spaces, I. J. Diff. Equat. **5**, 72—105 (1969).
2. Time-optimal control of solution of operational differential equations. J.S.I.A.M. Control, Ser. A **2**, 54—59 (1964).
3. Control in finite time of differential equations in Banach space. Comm. Pure Appl. Math. **19**, 17—34 (1966).

Fichera, G.

1. Sulla teoria generale dei problemi al contorno per le equazioni differenziali lineari. Rend. Acc. Lincei **21**, 46—55 and 166—172 (1956).
2. Linear elliptic equations of higher order in two independent variables and singular integral equations, with applications to anisotropic inhomogeneous elasitcity. Part. Diff. Equat. and Continuum Mech., University of Wisconsin Press (1961), 55—80.
3. On a unified theory of boundary value problems of elliptic-parabolic equations of second order, Boundary value Problems in Diff. Equat., University of Wisconsin Press 1960, 97—120.

4. Linear elliptic differential systems and eigenvalue problems, Lecture Notes No. 8, Berlin/Heidelberg/New York: Springer 1965.

Fishel, B.

1. Boundary value problems for second order, formally self-adjoint, elliptic differential equations. J. London Math. Soc. **33**, 62—70 (1958).

Foias, C.

1. Remarques sur les semi-groupes distributions d'opérateurs normaux. Portug. Math. **19**, 227—242 (1960).

Freeman, R. S.

1. On the spectrum and resolvent of homogeneous elliptic differential operators with constant coefficients. Bull. Amer. Math. Soc. **72**, 538—541 (1966).
2. On closed extensions of second order formally self-adjoint uniformly elliptic differential operators. Pacific J. of Math. **22**, 71—97 (1967).

Freeman, R. S., Bade, W. G.
Cf. Bade-Freeman.

Friberg, F.

1. Estimates for partially hypo-elliptic differential operators. Medd. Lund's Univ. Math. Sem. **17**, 1—97 (1963).

Friedman, A.

1. Generalized functions and partial differential equations, New York: Prentice-Hall 1963.
2. Partial differential equations of parabolic type, New York: Prentice-Hall 1964.
3. Uniqueness of solutions of ordinary differential inequalities in Hilbert space. Arch. Rat. Mech. Anal. **17**, 353—357 (1964).

Friedman, A., Shinbrot, M.

1. The initial value problem for the linearized equations of water waves. J. Math. Mech. **17**, 107—180 (1967).

Friedrichs, K. O.

1. On the differentiability of the solution of linear elliptic differential equations. Comm. Pure Appl. Math. **6**, 299—325 (1953).
2. Asymptotic phenomena in mathematical physics. Bull. Amer. Math. Soc. **61**, 485—504 (1955).
3. Symmetric positive linear differential equations. Comm. Pure Appl. Math. **11**, 333—418 (1958).
4. Spektraltheorie hallbeschränkter Operatoren. Math. Ann. **109**, 465—487; 685—713 (1934).

Friedrichs, K. O., Lax, P.

1. Boundary value problems for first order operators. Comm. Pure Appl. Math. **18**, 355—388 (1965).

Fujiwara, D.

1. A Characterization of exponential distribution semi-groups. J. Math. Soc. Japan **18**, 3, 267—274 (1966).
2. Concrete characterization of the domains of fractional powers of some elliptic differential operators of the second order. Proc. Japan Acad. **43**, 82—86 (1967).

Gamkrelidze, R., Boltyanskii, V., Mischenko, L., Pontryagin, L. S.
Cf. Pontryagin-Boltyanskii-Gamkrelidze-Mischenko.

Gårding, L.

1. Dirichlet's problem for linear elliptic partial differential equations. Math. Scand. **1**, 55—72 (1953).
2. On the asymptotic distribution of the eigenvalues and eigenfunctions of elliptic differential operators. Math. Scand. **1**, 237—255 (1953).
3. Energy inequalities for hyperbolic systems. Differential Analysis, Bombay Coll., London: Oxford University Press 1964, 209—225.

Gårding, L., Malgrange, B.

1. Opérateurs partiellement hypoelliptiques et partiellement elliptiques. Math. Scand. **9**, 5—21 (1961).

Garipov, R. M.

1. On the linear theory of gravity waves: the theorem of existence and uniqueness. Archive Rat. Mech. Anal. **24**, 352—362 (1967).

Garnir, H. G.

1. Les problèmes aux limites de la phsique mathématique, Basel: Birkhäuser 1958.

Gelfand, I. M.

1. Quelques problèmes d'analyse et d'équations différentielles. Uspehi Mat. Nauk **14**, No. 3, 3—19 (1959) [Amer. Math. Soc. Transl. (2) **26**, 201—218 (1963)].

Gelfand, I. M., Shilov, G. E.

1. Generalized Functions, Vol. I, II, III, Moscow 1958 (German translation: Berlin: Deutscher Verlag der Wissenschaften; French Translation: Paris: Dunod; English translation: New York: Academic Press).

Gelfand, I. M., Vilenkin, N. Ya.

1. Generalized Functions, Vol. IV, Moscow 1961 (English translation: New York: Academic Press).

Geymonat, G., Grisvard, P.

1. Problèmes aux limites elliptiques dans L^p. Orsay, Jan.—March, 1964.
2. Problemi ai limiti lineari ellittici negli spazi di Sobolev con peso. Le Matematiche **22**, 1—38 (1967).
3. Alcuni risultati di teoria spettrale per i problemi ai limiti lineari ellittici. Rend. Sem. Mat. Padova **38**, 121—173 (1967).

Giusti, E.

1. Equazioni quasi ellittiche e spazi $L^{p,\lambda}(\Omega, \delta)$, I and II. Ann. Mat. Pura Appl. **4**, LXXV, 313—354 (1967) and Ann. Sc. Norm. Sup. Pisa **21**, 353—372 (1967).

Giusti, E., Da-Prato, G.
Cf. Da Prato-Giusti.

Goldstein, J. A.

1. Abstract evolution equations. Trans. Amer. Math. Soc. **141**, 159—185 (1969).

Green, J. W.

1. An expansion method for parabolic partial differential operators. J. Res. Nat. Bur. Stand. **51**, 127—132 (1953).

Grisvard, P.

1. Espaces intermédiaires entre espaces de Sobolev avec poids. Ann. Sc. Norm. Sup. Pisa **17**, 255—296 (1963).
2. Identités entre espaces de traces. Math. Scand. **13**, 70—74 (1963).
3. Problèmes aux limites elliptiques dans L^2. Séminaire Lions-Schwartz, Paris, 1964—1965.
4. Commutativité de deux foncteurs d'interpolation et applications. Journal de Liouville **45**, 143—290 (1966).
5. Espace d'interpolation et équations opérationnelles. C. R. Acad. Sc. Paris **260**, 1536—1538 (1965).
6. Problèmes aux limites résolus par le calcul opérationnel. C. R. Acad. Sc. Paris **262**, 1306—1308 (1966).
7. Equations opérationnelles abstraites dans les espaces de Banach et problèmes aux limites dans des ouverts cylindriques. Ann. Sc. Norm. Pisa **21**, 307—347 (1967).
8. Caractérisation de quelques espaces d'interpolation. Ark. Rat. Mech. Anal. **25**, 40—63 (1967).
9. Equations differentielles abstraites. Cours Peccot, Collège de France, 1968; Ann. Sci. Ec. Norm. Sup. Paris (4) **2**, 311—395 (1969).

Grisvard, P., Baouendi, M. S.
Cf. Baouendi-Grisvard.

Grisvard, P., Geymonat, G.
Cf. Geymonat-Grisvard.

Grubb, G.

1. A characterization of the non local boundary value problems associated with an elliptic operator, July 1966, Department of Math., Stanford University. Ann. Sc. Norm. Sup. Pisa **22**, 425—513 (1968).

Guglielmino, F.

1. Sulla regolarizzazione delle soluzioni deboli dei problemi al contorno per operatori parabolici. Ricerche di Mat. **12**, 44—66 and 140—150 (1963).
2. Nuovi contributi allo studio delle equazioni paraboliche del secondo ordine di tipo non variazionale. Ricerche di Mat. **14**, 124—144 (1965).

Hadamard, J.

1. Lectures on Cauchy's problem in linear partial differential equations, New York: Dover 1952.

Hersh, R.

1. On Vibration, Diffusion or Equilibrium across a plane interface. Archive Rat. Mech. Anal. **21**, 5, 368—390 (1966).

Hille, E., Phillips, R. S.

1. Functional Analysis and Semi-groups, A.M.S. Coll. Pub., XXXI, 1957.

Hiraka, Y., Shiraiski, R.
Cf. Shiraiski-Hiraka.

Hörmander, L.

1. On the theory of general partial differential operators. Acta Math. **94**, 161 to 248 (1955).
2. On the interior regularity of the solutions of partial differential equations. Comm. Pure Appl. Math. **11**, 197—218 (1958).
3. On the regularity of the solutions of boundary problems. Acta Math. **99**, 225—264 (1958).
4. Definition of maximal differential operators. Ark. Mat. **3**, 501—504 (1958).
5. Weak and strong extensions of differential operators. Comm. Pure Appl. Math. **14**, 371—379 (1961).
6. Linear Partial differential Operators, Grundlehren math. Wissenschaften, Vol. 116, Berlin/Göttingen/Heidelberg: Springer 1963.
7. Pseudo-differential operators. Comm. Pure Appl. Math. **18**, 501—517 (1965).
8. Pseudo-differential operators and non elliptic boundary problems. Ann. of Math. **83**, 129—209 (1966).
9. Pseudo-differential operators and hypoelliptic equations. Proc. Symp. on singular integrals, Chicago, April 1966; Amer. Math. Soc. **10**, 138—183 (1967).
10. An introduction to complex analysis in several variables, Princeton: Van Nostrand 1966.
11. Hypoelliptic second order differential equations. Acta Math. **119**, 147—171 (1967).

Horvath, J.

1. Topological vector spaces and distributions, Readings, Mass.: Addison Wesley 1966.

Huet, D.

1. Phénomènes de perturbations singulières dans les problèmes aux limites. Ann. Inst. Fourier **10**, 1—96 (1960).
2. Phénomènes de perturbations singulières relatives au problème de Dirichlet dans un demi-espace. Ann. Sc. Norm. Sup. Pisa **18**, 425—448 (1964).
3. Sur quelques problèmes de perturbation singulière dans les espaces L_p, Revista Fac. Ciencias de Lisboa XI, 137—164 (1965).
4. Remarque sur un théorème d'Agmon et applications à quelques problèmes de perturbation singulière. Boll. U.M.I. **3**, 21, 219—227 (1966).
5. Perturbations singulières et régularité. C. R. Acad. Sc. Paris **265** A, 316 to 318 (1967).

Il'in, A. M., Kalashnikov, A. C., Oleinik, O. A.

1. Equations paraboliques linéaires du deuxième ordre. Uspehi Mat. Nauk **17**, 3—146 (1962) [Russian Math. Surv. **17**, 1—145 (1962)].

Il'in, V. A.

1. Résolution du problème mixte pour les équations hyperboliques et paraboliques. Uspehi Mat. Nauk **15**, 67—154 (1960) [Russian Math. Surv. **15**, 85—152 (1960)].

Itano, M.

1. On a trace theorem for the space $H^\mu(\mathbf{R}^N)$. J. Sci. Hiroshima Univ. **30**, 11—29 (1966).

Ito, S., Vamabe, H.

1. A unique continuation theorem for solutions of a parabolic differential equation. J. Math. Soc. Japan **10**, 314—321 (1958).

Jodeit Jr., M.

1. Symbols of Parabolic Singular Integrals, Proc. Symp. on singular integrals, Chicago, April 1966, Amer. Math. Soc. **10**, 184—195 (1967).

Jodeit Jr., M., Fabes E. F.
Cf. Fabes-Jodeit

John, F.

1. Plane waves and spherical means applied to partial differential equations, New York: Intersience Publishers 1955.

Jones Jr., B. F.

1. Singular integrals and parabolic equations. Bull. A.M.S. **69**, 501—503 (1963).
2. A class of singular integrals. Amer. J. Math. **86**, 441—462 (1964).
3. Singular integrals and a boundary value problem for the heat equations. Proc. Symp. on singular integrals, Chicago, April 1966, Amer. Math. Soc. **10**, 196—207 (1967).
4. Lipschitz spaces and the heat equation. J. Math. Mech. **18**, 379—409 (1969).

Jorgens, K.

1. An asymptotic expansion in the theory of neutron transport. Comm. Pure Appl. Math. **11**, 291—242 (1958).

Juberg, R. K.

1. A generalization of a result of Tihonov. Archive for Rat. Mech. Anal. **22**, 153—162 (1966).
2. General mixed problems on a half-space for certain higher order parabolic equations in one space variable. Ann. Sc. Norm. Sup. Pisa **20**, 243—264 (1966).

Kadlec, J.

1. Solution du premier problème aux limites pour une généralisation de l'équation de la chaleur dans des classes de fonctions avec dérivée fractionnaire par rapport au temps. Cecoslovack Mat. J. **16**, 91—113 (1966).

Kalashnikov, A. C., Il'in, A. M., Oleinik, O. A.
Cf. Il'in-Kalashnikov-Oleinik.

Kamin, L. I., Maslennikova, V. N.

1. Inégalité à la frontière du type de Schauder pour les problèmes de dérivées obliques pour équations paraboliques dans des domaines non cylindriques. Sibirsk Mat. J. **7**, 83–128 (1966).

Kaplan, S.

1. Abstract boundary value problems for linear parabolic equations. Ann. Sc. Norm. Sup. Pisa **20**, 395–420 (1966).

Kato, T.

1. Integration of the equation of evolution in a Banach space. Journal Mat. Soc. Japan **5**, 208–234 (1953).
2. Abstract evolution equations of parabolic type in Banach and Hilbert spaces. Nagoya Math. Journ. **19**, 93–125 (1961).
3. Fractional powers of dissipative operators. Journ. Mat. Soc. Japan **13**, 246–274 (1961).
4. A Generalization of the Heinz inequality. Proc. Japan Acad. **6**, 304–308 (1961).
5. Perturbation theory for linear operators, Grundlehren math. Wissenschaften, Vol. 132, Berlin/Heidelberg/New York: Springer 1966.

Kato, T., Tanabe, H.

1. On the evolution equation. Osaka Math. Journal **14**, 107–133 (1962).

Kohn, J. J., Nirenberg, L.

1. Non coercive boundary value problems. Comm. Pure Appl. Math. **18**, 443 to 492 (1965).
2. On algebra of pseudo-differential operators. Comm. Pure Appl. Math. **18**, 296–305 (1965).
3. Degenerate elliptic-parabolic equations of second order. Comm. Pure Appl. Math. **20**, 797–872 (1967).

Kostiouchenko, A. G., Levitan, V. M.

1. Sur le comportement asymptotique des valeurs propres des problèmes opérationnels du type Sturm-Liouville. Analyse Fonct. et Applic. **1**, 1, 86–96 (1967).

Krein, M. G.

1. Théorie des prolongements auto-adjoints d'opérateurs symétriques semi-bornés et applications. Mat. Sbornik **20** (62), 431–495 (1947).

Krein, S. G.

1. Sur quelques classes de problèmes aux limites bien posés. Dokl. Ak. Nauk **114**, 1162–1165 (1957).
2. Sur un théorème d'interpolation dans la théorie des opérateurs. Dokl. Ak. Nauk **130**, 491–494 (1960) [Soviet Math. **1**, 61–64 (1960)].
3. Sur la notion d'échelle normale d'espace. Dokl. Ak. Nauk **132**, 510–513 (1960) [Soviet Math. **1**, 586–589 (1960)].

4. Linear Differential Equations in Banach spaces, Moscow 1967.
5. Problème de Cauchy bien posé et analyticité de la solution de l'équation d'évolution. Dokl. Ak. Nauk **171**, 1033—1036 (1966) [Soviet Math. **7**, 1594 to 1598 (1966)].

Krein, S. G., Laptiev, G.

1. Boundary value problems for second order differential equations in Banach spaces, I, II. Diff. Equations, II **3**, 382—390; **7** 919—926 (1966).

Kroujkov, S. N.

1. Inégalités à priori et propriétés des solutions d'équations elliptiques et paraboliques. Mat. Sbornik **65** (107), 522—570 (1964).
2. Equations paraboliques non linéaires à deux variables indépendantes. Troudi Mosk. Mat. Obv. **16**, 329—346 (1967).

Kroujkov, S. N., Oleinik, O. A.
Cf. Oleinik-Kroujkov.

Krzyzanski, M. Schauder, J.

1. Quasilineare Differentialgleichungen zweiter Ordnung von hyperbolischen Typus, Gemischte Randwertaufgaben. Studia Math. **6**, 162—189 (1936).

Ladyženskaya, O. A.

1. Problèmes mixtes pour les équations hyperboliques, Moscow 1953.
2. Sur la solution d'équations opérationnelles non stationnaires. Mat. Sbornik **39**, 491—524 (1956).
3. Sur les équations opérationnelles non stationnaires et leurs applications aux problèmes linéaires de la physique mathématique. Mat. Sbornik **45**, 123 to 158 (1958).

Ladyženskaya, O. A., Solonnikov, V. A., Uraltzeva, N. N.

1. Equations de types paraboliques linéaires et quasi-linéaires, Moscow 1967.

Ladyženskaya, O. A., Uraltzeva, N. N.

1. Equations elliptiques linéaires et quasi-linéaires, Moscow 1964.
2. Un problème aux limites pour équations paraboliques linéaires et quasilinéaires. Izsvestija Akad. Nauk **26**, 5—52 (1962) [Amer. Math. Soc. Transl. (2) **47**, 217—267 (1965)].

Ladyženskaya, O. A., Vishik, I. M.
Cf. Vishik-Ladyženskaya.

Landis, E. M.

1. Sur certaines propriétés qualitatives dans la théorie des équations elliptiques et paraboliques. Uspehi Mat. Nauk **14**, 21—85 (1959) [Amer. Math. Soc. Transl. (2) **20**, 173—238 (1962)].

Laptiev, G., Krein, S. G.
Cf. Krein-Laptiev.

Lax, P. D.

1. On Cauchy's problem for hyperbolic equations and the differentiability of the solutions of elliptic equations. Comm. Pure Applied Math **8**, 615—653 (1955).

2. A Phragmen-Lindelöf theorem in harmonic analysis and its applications to some questions in the theory of elliptic equations. Comm. Pure Applied Math. **10**, 361—389 (1957).

Lax, P. D., Friedrichs, K. O.
Cf. Friedrichs-Lax.

Lax, P. D., Milgram, N.

1. Parabolic equations. Contributions to the theory of partial differential equations, Annals of Math. Studies, No. 33, Princeton, 1954, 167—190.

Lax, P. D., Phillips, R. S.

1. Local boundary conditions for dissipative symmetric linear differential operators. Comm. Pure Appl. Math. **13**, 427—455 (1960).
2. Scattering Theory, New York: Academic Press 1967.

Leray, J.

1. Hyperbolic differential equations, Princeton 1952.
2. Etude de diverses équations intégrales non linéaires et de quelques problèmes que pose l'hydrodynamique. J. Math. pures appl. **12**, 1—82 (1933).
3. Essai sur les mouvements plans d'un liquide visqueux que limitent les parois. J. Math. pures appl. **13**, 331—418 (1934).
4. Sur les mouvements d'un liquide visqueux emplissant l'espace. Acta Math. **63**, 193—248 (1934).

Levitan, V. M.

1. Book to appear.

Levitan, V. M., Kostiouchenko, A. G.
Cf. Kostiouchenko-Levitan.

Lions, J. L.

1. Supports dans la transformation de Laplace. J. d'Analyse Math. d'Israël **2**, 369—380 (1952—1953).
2. Problèmes aux limites en théorie des distributions. Acta Math. **94**, 13—153 (1955).
3. Sur quelques problèmes aux limites relatifs à des opérateurs différentiels elliptiques. Bull. Soc. Math. France **83**, 225—250 (1955).
4. Some questions on elliptic equations. Tata Inst. of Fundamental Research, Bombay, 1957.
5. Sur les problèmes mixtes pour certains systèmes paraboliques dans des ouverts non cylindriques. Ann. Inst. Fourier **7**, 143—182 (1957).
6. Boundary value problems, Technical Report No.. 13, University of Kansas, Lawrence, 1957.
7. Une remarque sur les applications du théorème de Hille-Yosida. J. Math. Soc. Japan **9**, 62—70 (1957).
8. Problèmes mixtes abstraits. Proc. Int. Congress of Math., Edinburgh, 1958, 389—397.
9. Espaces intermédiaires entre espaces hilbertiens et applications. Bull. Math. Soc. Sc. Math. Phys. R. P. Roumaine **2**, **50**, 419—432 (1958).
10. Sur les problèmes aux limites du type dérivées obliques. Ann. of Math. **64**, 207—239 (1959).

11. Théorèmes de traces et d'interpolation (I) ... (V); (I), (II) Ann. Sc. Norm. Sup. Pisa **13**, 389—403 (1959); **14**, 317—331 (1960); (III) Journal de Liouville **42**, 196—203 (1963); (IV) Math. Annalen **151**, 42—56 (1963); (V) Anais de Acad. Brasileira de Ciencias **35**, 1—10 (1963).
12. Les semi-groupes distributions. Port. Math. **19**, 141—164 (1960).
13. Equations différentielles opérationnelles et problèmes aux limites, Grundlehren, Vol. 111, Berlin/Göttingen/Heidelberg: Springer 1961.
14. Sur les espaces d'interpolation; dualité. Math. Scand. **9**, 147—177 (1961).
15. Une construction d'espace d'interpolation. C. R. Acad. Sc. Paris **251**, 1853 to 1855 (1961).
16. On a trace problem. Rend. Sem. Mat. Padova **31**, 232—242 (1961).
17. Espaces d'interpolation et domaines de puissances fractionnaires d'opérateurs. J. Math. Soc. Japan **14**, 233—241 (1962).
18. Remarques sur les espaces d'interpolation et les problèmes aux limites. Colloque C.N.R.S., No. 117, "Les équations aux dérivées partielles", Paris, 75—86 June 1962.
19. Properties of some interpolation spaces. J. Math. and Mech. **11**, 969—978 (1962).
20. Dérivées intermédiaires et espaces intermédiaires. C. R. Acad. Sc. Paris **256**, 4343—4345 (1963).
21. Equations différentielles opérationnelles dans les espaces de Hilbert, Centro Int. Mat. Estivo, Varenna, 1963 (Equazioni differenziali astratte, Roma: Cremonese 1963).
22. Remarques sur les équations différentielles opérationnelles. Osaka Math. Journal **15**, 131—142 (1963).
23. Quelques remarques sur les équations différentielles opérationnelles du premier ordre. Rend. Sem. Mat. Padova **33**, 213—225 (1963).
24. On some optimization problems for linear parabolic equations, Functional Analysis and Optimization, New York: Academic Press 1966.
25. Optimisation pour certaines classes d'équations d'évolution non linéaires. Ann. Mat. Pura Appl. **4**, LXXII, 275—294 (1966).
26. Sur le contrôle optimal de systèmes décrits par des équations aux dérivées partielles linéaires, (I) Remarques générales, (II) Equations elliptiques, (III) Equations d'évolution. C. R. Acad. Sc. Paris **263**,661—663; 713—715; 776—779 (1966).
27. Remarks on evolution inequalities. J. Math. Soc. Japan **18**, 331—342 (1966).
28. Control problems in systems described by partial differential equations, in Mathematical Theory of Control, edited by Balakrishnan and Neustadt, New York: Academic Press 1967, 251—271.
29. Problèmes aux limites non homogènes à données irrégulières; une méthode d'approximation, Cours C.I.M.E., Ispra, July 1967, Roma: Cremonese 1968.
30. Contrôle optimal de systèmes gouvernés par des équations aux dérivées partielles, Paris: Dunod 1968.
31. Une remarque sur les problèmes d'évolution non linéaires dans des domaines non cylindriques. Revue Roumaine Math. Pures et Appliquées XI, 11—18 (1964).

Lions, J. L., Brezis, H.
Cf. Brezis-Lions.

Lions, J. L., Magenes, E.

1. Problèmes aux limites non homogènes(I) ... (VII); (I), (III), (IV), (V) Ann. Sc. Norm. Sup. Pisa **14**, 259—308 (1960); **15**, 39—101; 311—326 (1961); **16**,

1—44 (1962); (II) Ann. Inst. Fourier **11**, 137—178 (1961); (VI), Journal d'Analyse Math. **11**, 165—188 (1963); (VII), Ann.Mat. Pura Appl. **4**, LXIII, 201—224 (1963).
2. Remarques sur les problèmes aux limites pour opérateurs paraboliques. C. R. Acad. Sc. Paris **251**, 2118—2120 (1960).
3. Remarques sur les problèmes aux limites linéaires elliptiques. Rend. Acc. Naz. Lincei **8**, 32, 873—883 (1962).

Lions, J. L., Malgrange, B.

1. Sur l'unicité rétrograde. Math. Scand. **8**, 277—286 (1960).

Lions, J. L., Peetre, J.

1. Sur une classe d'espaces d'interpolation, Institut des Hautes Etudes Scientifiques. Publ. Math. **19**, 5—68 (1964).

Lions, J. L., Raviart, P. A.

1. Remarques sur la résolution et l'approximation d'équations d'évolutions couplées. I.C.C. Bulletin **5**, 1—20 (1966).
2. Remarques sur la résolution, exacte et approchée, d'équations d'évolution paraboliques à coefficients, opérateurs non bornés. Calcolo **4**, 221—234 (1967).

Lions, J. L., Stampacchia, G.

1. Variational inequalities. Comm. Pure Applied Math. **20**, 493—519 (1967). [Development of the note in C. R. Acad. Sc. Paris **261**, 25—27 (1965)].

Lipko, B., Eidelman, S. D.

1. Contribution à la théorie des potentiels paraboliques. Dokl. Ak. Nauk **166**, 1050—1053 (1966) [Soviet Math. **7**, 237—240 (1966)].

Littmann, W.

1. The wave operator and L^p norms. J. Math. Mech. **12**, 55—68 (1963).

Lizorkin, P. S., Nikolski, S. M.

1. Classification des fonctions différentiables par des espaces avec dérivées mixtes dominantes. Trudy Mat. Inst. Steklov **77** (1965).

Magenes, E.

1. Sul problema di Dirichlet per le equazioni lineari ellittiche in due variabili. Ann. Mat. Pura Appl. **4**, 48, 257—279 (1959).
2. Sur les problèmes aux limites pour les équations linéaires elliptiques. Colloque C.N.R.S., No. 117, Les équations aux dérivées partielles, Paris, 95—111 (1962).
3. Spazi di interpolazione ed equazioni a derivate parziali. Atti VII Congresso U.M.I., Genova 1963, Roma: Cremonese 1964, 134—197.

Magenes, E., Lions, J. L.
Cf. Lions-Magenes.

Malgrange, B., Gårding, L.
Cf. Gårding-Malgrange.

Malgrange, B., Lions, J. L.
Cf. Lions-Malgrange.

Maslennikova, V. N., Kaminin, L. I.
Cf. Kaminin-Maslennikova.

Martyniuk, A. E.

1. Certaines nouvelles applications des méthodes du type Galerkin. Mat. Sbornik **49**, 85—108 (1959).

Matsuzawa, T.

1. Regularity at the boundary for solution of hypo-elliptic equations. Osaka J. Math. **3**, 313—334 (1966).
2. On quasi-elliptic boundary problems. Trans. Amer. Math. Soc. **133**, 241—265 (1968).

Milgram, N., Lax, P. D.
Cf. Lax-Milgram.

Milgram, A. N., Aronszajn, N.
Cf. Aronszajn-Milgram.

Mischenko, E., Pontryagin, L. S., Boltyanskii, V., Gamkrelidze, R.
Cf. Pontryagin-Boltyanskii-Gamkrelidze-Mischenko.

Mizohata, S.

1. Quelques problèmes au bord, de type mixte, pour des équations hyperboliques, Seminaire Leray, Collège de France, Paris, December 1966.
2. Hypoellipticité des équations paraboliques. Bull. Soc. Math. France **85**, 15—50 (1957).
3. Systèmes hyperboliques. J. Math. Soc. Japan **11**, 205—233 (1959).
4. Lectures on the Cauchy problem, Tata Inst. of Fundamental Research, Bombay, 1965.
5. Unicité du prolongement des solutions pour quelques opérateurs différentiels paraboliques. Memoirs Univ. of Tokyo XXXI, 219—239 (1958).

Morrey, C. B., Nirenberg, L.

1. On the analyticity of the solution of linear elliptic systems of partial differential equations. Comm. Pure Applied Math. **10**, 271—290 (1957).

Mosco, M., Da-Prato, G.
Cf. Da-Prato Mosco.

Moser, J.

1. A Harnack inequality for parabolic differential equations. Comm. Pure Appl. Math. **17**, 101—134 (1964).

Nash, J.

1. Continuity of the solution of parabolic and elliptic equations. Amer. J. Math. **80**, 931—954 (1958).

Nelson, E.

1. L'équation de Schrödinger et les intégrales de Feynman. Colloque C.N.R.S., No. 117, Les équations aux dérivées partielles, Paris, June 1962, 151—158.

Nikolski, S. M.

1. Théorèmes d'inclusion, de prolongement et d'approximation pour les fonctions différentiables de plusieurs variables. Uspehi Mat. Nauk **16**, No. 5, 63—114 (1961) [Russian Math. Surv. **16**, No. 5, 45—104 (1916)].
2. Le premier problème aux limites pour une équation linéaire générale. Dokl. Ak. Nauk **146**, 767—769 (1962) [Soviet Math. **3**, 1388—1390 (1962)].
3. Sur les propriétés à la frontière des fonctions de plusieurs variables. Dokl. Ak. Nauk **146**, 542—546 (1962) [Soviet Math. **3**, 1357—1360 (1962)].
4. Une représentation constructive de la classe zéro des fonctions différentiables de plusieurs variables. Dokl. Ak. Nauk **170**, 512—515 [Soviet Math. **7**, 1227—1230 (1966)].
5. Théorie de l'approximation des fonctions différentiables de plusieurs variables et théorèmes de plongement, Moscow, 1968.

Nikolskii, S. M., Lizorkin, P. I.
Cf. Lizorkin-Nikolskii.

Nirenberg, L., Agmon, S.
Cj. Agmon-Nirenberg.

Nirenberg, L., Kohn, J.
Cf. Kohn-Nirenberg.

Odhnoff, J.

1. Un exemple de non-unicité d'une équation différentielle opérationnelle. C. R. Acad. Sci. Paris **258**, 1689—1691 (1964).
2. Operators generated by differential problems with eigenvalue parameter in equation and boundary conditions. Medd. Lunds Univ. Mat. Sem. **14**, 3—80 (1959).

Oleinik, O. A.

1. Sur un problème de G. Fichera. Dokl. Ak. Nauk **157**, 1297—1300 [Soviet Math. **5**, 1129—1133 (1964)].
2. Alcuni risultati sulle equazioni lineari e quasi lineari ellittico-paraboliche a derivate parziali del secondo ordine. Accad. Naz. Lincei (8) XL, 775—784 (1966).
3. Sur les équations linéaires du deuxième ordre à forme caractéristique non négative. Mat. Sbornik **69**, 111—140 (1966).
4. Problèmes de Cauchy et aux limites pour équations hyperboliques du deuxième ordre dégénérant dans le domaine et sur la frontière. Dokl. Ak. Nauk **169**, 525—528 (1966) [Soviet Math. **7**, 969—973 (1966)].
5. Sur les équations de type elliptique qui dégénérent à la frontière. Dokl. Ak. Nauk **87** 885—888 (1952).
6. Sur les équations de type elliptique avec un petit paramètre dans les dérivées d'ordre maximum. Mat. Sbornik **31**, 104—117 (1952).

Oleinik, O. A., Il'in, A. M., Kalashnikov, A. C.
Cf. Kalashnikov-Il'in-Oleinik.

Oleinik, O. A., Kroujkov, S. N.

1. Equations paraboliques du 2eme ordre à plusieurs variables indépendantes. Uspehi Mat. Nauk **16**, 115—155 (1961).

Pagni, M.

1. Problemi al contorno per una certa classe di equazioni lineari alle derivate parziali. Atti Sem. Mat. Fis. Modena **13**, 119—164 (1964).
2. Un teorema di tracce. Rend. Acc. Naz. Lincei VII, **38**, 627—631 (1965).
3. Problemi con condizioni al contorno omogenee per una certa classe di operatori formalmente ipoellittici. Ann. Mat. Pura Appl. **4**, LXXII, 201—212 (1966).
4. Su un problema di tracce. Atti Sem. Mat. Fis. Modena **15**, 175—181 (1966).

Panejach, B. P., Volevich, L. R.
Cf. Volevich-Panejach.

Peetre, J.

1. Théorèmes de régularité pour quelques classes d'opérateurs différentiels. Med. Lund Univ. Mat. Sem. **16**, 1—122 (1959).
2. Another approach to elliptic boundary problems. Comm. Pure Appl. Math. **14**, 711—731 (1961).
3. Mixed problems for higher order elliptic equations in two variables, I. Ann. Sc. Norm. Sup. Pisa **15**, 337—353 (1961).
4. Elliptic partial Differential equations of higher order, Lect. Series No. 40, Inst. for Fluid Dynamics and App. Math., University of Maryland, 1962.
5. Nouvelles propriétés d'espaces d'interpolation. C. R. Acad. Sc. Paris **256**, 54—55 (1963).
6. A theory of interpolation of normed spaces, Lectures Brasilia, 1963 (Notas de Matematica, No. 39, 1968).
7. On the theory of interpolation spaces. Revista Un. Mat. Argentina **23**, 49 to 66 (1967).
8. Espaces d'interpolation, généralisations, applications. Rend Sem. Mat. Fis. Milano **34**, 133—161 (1964).
9. Sur la théorie des semi-groupes distributions. Sem. Leray, Collège de France, Nov. 1963, May 1964, 79—98.
10. On an interpolation theorem of Foias and Lions. Acta Szeged **25**, 255—261 (1964).
11. Some remarks on continuous orthogonal expansions, and eigenfunction expansions for positive self-adjoint elliptic operators with variable coefficients. Math. Scand. **17**, 56—64 (1965).
12. Espaces d'interpolation et théorème de Sobolev. Ann. Inst. Fourier **16**, 279—317 (1966).
13. A formula for the solution of Dirichlet's problem (to be puslished in Summa Brasil. Math.).

Peetre, J., Lions, J. L.
Cf. Lions-Peetre.

Petrowski, J. G.

1. Sur l'analyticité des solutions des systèmes d'équations différentielles. Mat. Sbornik **5**, 47, 3—70 (1939).
2. Über das Cauchysche Problem für Systeme von partiellen Differentialgleichungen. Mat. Sbornik **2**, 44, 815—870 (1937).

Peyser, G.

1. Energy integrals for the mixed problem in hyperbolic partial differential equations of higher order. J. Rat. Mech. Anal. **6**, 641—669 (1957).

Phillips, R. S.

1. Dissipative operators and hyperbolic systems of partial differential equations. Trans. Amer. Math. Soc. **90**, 123–254 (1959).
2. Semi-groups of contraction operators, C.I.M.E., Roma: Cremonese **1963**, 1–48.

Phillips, R. S., Hille, E.
Cf. Hille-Phillips.

Phillips, R. S., Lax, P.
Cf. Lax-Phillips.

Phillips, R. S., Sarason, L.

1. Singular Symmetric Positive First Order Differential Operators. J. of Math. and Mech. **15**, 253–271 (1966).

Pini, B.

1. Un problema di valori al contorno generalizzato per la equazione lineare a derivate parziali lineare parabolica del secondo ordine. Riv. Mat. Univ. Parma **3**, 153–187 (1952).
2. Sui sistemi di equazioni lineari a derivate parziali del second ordine dei tipi ellittico e parabolico. Rend. Sem. Mat. Padova **22**, 265–280 (1953).
3. Sull'unicità della soluzione del problema di Dirichlet per le equazioni lineari ellittiche in due variabili. Rend. Sem. Mat. Padova **26**, 223–231 (1956).
4. Sul problema di Dirichlet per le equazioni a derivate parziali lineari ellittiche in due variabili. Rend. Sem. Mat. Padova **26**, 177–200 (1956).
5. Sul problema fondamentale di valori al contorno per una classe di equazioni paraboliche lineari. Ann. Mat. Pura Appl. **4**, XLIII, 261–297 (1957).
6. Sulle equazioni lineari pseudoparaboliche, I and II. Rend. Sem. Mat. Padova **30**, 255–280 and 361–375 (1960).
7. Sui sistemi lineari pseudoparabolici, Ricerche di Mat. **10**, 33–65 (1961).
8. Proprietà locali delle soluzioni di una classe di equazioni ipoellittiche. Rend. Sem. Mat. Padova **32**, 221–238 (1962).
9. Su un problema al contorno per certe equazioni ipoellittiche. Revue Roumaine Math. pures appl. **9** 7, 643–653 (1964).
10. Sulle tracce di un certo spazio funzionale, I, II. Rend. Acc. Naz. Lincei VIII, **37**, 28–34 (1964); VIII, **38**, 1–6 (1965).
11. Su un problema tipico relativo a una certa classe di euqazioni ipoellittiche. Atti Acc. sc. Bologna **12** 1, 1–26 (1964).
12. Proprietà al contorno delle funzioni di classe $H^{p_1,\ldots,p_n}$ per regioni dotate di punti angolosi. Ann. Mat. Pura Appl. **4** 73, 33–54 (1966).
13. Sulla regolarità e irregolarità della frontiera per il primo problema di valori al contorno relativo all'equazione del calore. Ann. Mat. Pura Appl. **40**, 69 to 88 (1955).
14. Sui punti singolari delle soluzioni delle equazioni paraboliche lineari. Ann. Univ. Ferrara **2**, 2–12 (1953).

Plejel, A.

1. Green's function and asymptotic distribution of eigenvalues and eigenfunctions. Proc. Symp. Spectral Theory and Differential Problems, Stillwater, Okl., **1951**, 439–454.

Pogorelenko, V. A., Sobolevski, P. E.

1. Equations hyperboliques dans des espaces de Banach. Uspehi Mat. Nauk **22** 133, 170—172 (1967).

Pontryagin, L. S., Boltyanskii, V., Gamkrelidze, R., Mischenko, E.

1. The Mathematical Theory of Optimal Processes, New York: Interscience Publishers 1962.

Poulsen, E. T.

1. Evolutionsgleichungen in Banach-Räumen. Math. Zeit. **90**, 286—309 (1965).

Pozzi, G. A.

1. Problemi di Cauchy e problemi ai limiti per equazione di evoluzione del tipo di Schrödinger lineari e non lineari I, II, Ann. Mat. Pura Appl., **4**, 78 197—258 (1968); **4**, 81, 205—248 (1969).

Prouse, G.

1. Analisi di alcuni classici problemi di propagazione. Rend. Sem. Mat. Padova **32**, 338—373 (1962).

Prouse, G., Amerio, L.
Cf. Amerio-Prouse.

Ramazanov, M. D.

1. Sur une classe d'espaces fonctionnels, Differenzialne Uravnienia, II, No. 1 65—82 (1966).
2. A priori estimates of solutions of the Dirichlet problem for certain hypoelliptic equations. Dokl. Akad. Nauk **179**, 783—785 (1968).

Raksin, V. G., Sobolevski, P. E.

1. Problème de Cauchy pour les équations différentielles du deuxième ordre dans les espaces de Banach. Sibirsk. Mat. J. **8**, 70—90 (1967).

Raviart, P. A., Lions, J. L.
Cf. Lions-Raviart.

Rivière, N. M., Fabes, E. B.
Cf. Fabes-Rivière.

Robinson, A., Campbell, L.
Cf. Campbell-Robinson.

Rogak, E. D.

1. A mixed problem for the wave equation in a time dependent domain. Archive Rat. Mech. Anal. **22**, 24—36 (1966).

Russel, D. L.

1. Optimal regulation of linear symmetric hyperbolic systems with finite dimensional control. J. Siam Control **4**, 276—294 (1966).

Sarason, L.

1. Differentiable solution of symmetrizable and singular symmetric first order systems. Archive for Rat. Mech. and Anal. **26**, 357—384 (1967).
2. Elliptic regularization for symmetric positive systems. J. Math. Mech. **16**, 807—827 (1967).

Sarason, L., Phillips, R. S.
Cf. Phillips-Sarason.

Scarpini, F.

1. Su taluni problemi al contorno relativi ad una equazione alle derivate parziali del terzo ordine. Rend. Circ. Mat. Palermo **14**, 5—33 (1965).

Schauder, I.

1. Das Anfangswertproblem einer quasilinearen hyperbolischen Differentialgleichung zweiter Ordnung in beliebiger Anzahl von unabhängigen Veränderlichen. Fund. Math. **24**, 213—246 (1935).

Schauder, I., Krzyzanski, M.
Cf. Krzyzanski-Schauder.

Schechter, M.

1. Non local elliptic Boundary value Problems. Ann. Sc. Norm. Sup. Pisa **20**, 421—442 (1966).

Schmaedeke, W.

1. Mathematical theory of optimal control for semi-linear hyperbolic systems in two independent variables. J.S.I.A.M. Control **5**, No. 1, 138—152 (1967).

Schwartz, J., Dunford, N.
Cf. Dunford-Schwartz.

Schwartz, L.

1. Théorie des distributions, I and II, Paris: Hermann 1950—1951 (second edition 1957).
2. Transformation de Laplace des distributions. Sem. Math. Univ. Lund, volume dedicated to M. Riesz, 1952, 196—206.
3. Les équations d'évolution liées au produit de convolution. Ann. Inst. Fourier **2**, 19—49 (1952).
4. Les travaux de Gårding sur le problème de Dirichlet. Sem. Bourbaki, Paris, May 1952.
5. Espaces de fonctions différentiables à valeurs vectorielles. J. d'Analyse Math. **4**, 88—148 (1954—1955).
6. Distributions à valeurs vectorielles, I and II. Ann. Inst. Fourierr **7**, 1—141 (1957); **8**, 1—209 (1958).
7. Su alcuni problemi della teoria delle equazioni differenziali lineari di tipo ellittico. Rend. Sem. Mat. Fis. Milano **27**, 211—249 (1958).

Serrin, J., Aronson, D. G.
Cf. Aronson-Serrin.

Shilov, G. E., Gelfand, I. M.
Cf. Gelfand-Shilov.

Shinbrot, M., Friedman, A.
Cf. Friedman-Shinbrot.

Shiraishi, R., Hiraka, Y.

1. Convolution maps and semi-group distributions. J. Sci. Hiroshima University **28** 1, 71—88 (1964).

Slobodetski, L. N.

1. Espaces de Sobolev généralisés et applications aux problèmes aux limites pour les équations différentielles aux dérivées partielles. Uchen. Zap. Leningrad Gos. Ped. Inst. **197**, 54—112 (1958).
2. Estimations des solutions des systèmes elliptiques et paraboliques. Dokl. Akad. Nauk **120**, 468—471 (1958).
3. Estimations dans L_2 des solutions des systèmes linéaires elliptiques et paraboliques. Vestink Lenin. Gos. Univ. **7**, 28—47 (1960).

Sobolev, S. L.

1. Sur les problèmes aux limites pour les équations polyharmoniques. Mat. Sbornik **2** 44, 467—500 (1937).
2. Applications de l'Analyse fonctionnelle aux équations de la Physique Mathématique, Leningrad 1950 (German translation: Akademie-Verlag Berlin: English translation: Amer. Math. Soc.).
3. Méthode nouvelle à résoudre le problème de Cauchy pour les équations hyperboliques normales. Mat. Sbornik **43**, 39—71 (1936).

Sobolev, S. L., Vishik, I. M.

1. Nouvelle formulation générale des problèmes aux limites. Dokl. Akad. Nauk **111**, 521—523 (1956).

Sobolevski, P. E.

1. Sur la fonction de Green des puissances fractionnaires (et en particulier entières) des opérateurs elliptiques. Dokl. Akad. Nauk **142**, 804—807 (1962) [Soviet Math. **3**, 183—187 (1962)].
2. Inégalités de coercivité pour les équations paraboliques abstraites. Dokl. Akad. Nauk **157**, 52—55 (1964) [Soviet Math. **5**, 894—897 (1964)].
3. Solutions généralisées des équations différentielles du 1er ordre dans les espaces de Banach. Dokl. Akad. Nauk **165**, 486—489 (1965) [Soviet Math. **6**, 1465—1468 (1965)].
4. Sur la solution d'équations paraboliques dans les espaces de Lorentz. Uspehi Mat. Nauk **21**, (127) 169—171 (1966).
5. Sur les puissances fractionnaires d'opérateurs faiblement positifs. Dokl. Akad. Nauk **166**, 1296—1299 (1966) [Soviet Math. **7**, 287—291 (1966)].
6. Sur les équations de type parabolique dans les espaces de Banach. Trudy Moskov Mat. Obs. **10**, 297—350 (1961) [Amer. Math. Soc. Transl. (2) **49**, 1—62 (1966)].

Sobolevski, P. E., Pogorelenko, V. A.
Cf. Pogorelenko-Sobolevski.

Sobolevski, P. E., Raskin, V. R.
Cf. Raskin-Sobolevski.

Solonnikov, V. A.

1. Sur les prolèmes aux limites généraux au sens de Douglis et Nirenberg. (I) Isvestia Akad. Nauk **28**, 665—706 (1964); (II) Trudy Steklov XCII, 233—297 (1966) [Amer. Math. Soc. Transl. (2) **56**, 193—231 (1966)].
2. Inégalité à priori pour les équations du deuxième ordre de type parabolique. Trudy Steklov LXX, 133—212 (1964).
3. Sur les propriétés différentielles de la solution du premier problème aux limites pour les systèmes non stationnaires de Navier-Stokes. Trudy Steklov LXXIII, 221—291 (1964).
4. Estimations a priori pour le système non stationnaire linéarisé de l'équation de Navier-Stokes. Trudy Steklov **70**, 213—317 (1964).

Solonnikov, V. A., Ladyženskaya, O. A., Uraltzeva, N. N.
Cf. Ladyženskaya-Solonnikov-Uraltzeva.

Stampacchia, G.

1. Formes bilinéaires coercitives sur les ensembles convexes. C. R. Acad. Sc. Paris **258**, 4413—4416 (1964).
2. Le problème de Dirichlet pour les équations elliptiques du second ordre à coefficients discontinus. Ann. Inst. Fourier **15**, 189—258 (1965).
3. The $\mathscr{L}^{p,\lambda}$ spaces and applications to the theory of partial differential equations. Proceedings of the Colloquium "Equa. diff. II", Bratislava, September 1966.
4. Su un problema relativo alle equazioni di tipo ellittico del secondo ordine. Ricerche Mat. **5**, 3—24 (1956).
5. Equations elliptiques à données discontinues, Séminaire Schwartz, Paris, **5**, No. 4 (1960—1961).
6. Sistemi di equazioni di tipo ellittico a derivate parziali del primo ordine e proprietà delle estremali degli integrali multipli. Ricerche Mat. **1**, 200—226 (1952).

Stampacchia, G., Lions, J. L.
Cf. Lions-Stampacchia.

Strauss, W.

1. On continuity of functions with values in various Banach spaces. Pacific J. of Math. **19**, 543—551 (1966).

Tacklind, S.

1. Sur les classes quasi analytiques des solutions des équations aux dérivées partielles paraboliques. Nova Acta Soc. Sc. Upsahensis (4) **10**, 1—57 (1936).

Tanabe, H.

1. On the equations of evolution in a Banach space. Osaka Math. J. **11**, 121 to 145 (1959); **12**, 145—166 and 363—376 (1960).
2. Note on uniqueness of solution of differential inequalities of parabolic type. Osaka Math. J. **2**, 191—204 (1965).
3. On estimates for derivatives of solutions of weighted elliptic boundary value problems. Osaka Math. J. **3**, 163—182 (1966).

Tanabe, H., Kato, T.
Cf. Kato-Tanabe.

Thomee, V.

1. Existence proofs for mixed problems for hyperbolic differential equations in two independent variables by means of the continuity methods. Math. Scand. **6**, 5—32 (1958).

Tikonov, A. N.

1. Théorèmes d'unicité pour l'équation de la chaleur. Mat. Sbornik **42**, 199 to 216 (1935).
2. Sur les conditions aux limites qui contiennent des dérivées d'ordre plus grand que l'ordre de l'équation. Mat. Sbornik **26**, 38, 35—56 (1950) [Amer. Mat. Soc. Transl. s. **1**, No. **4**, 440—465 (1962)].

Torelli, G.

1. Un complemento ad un teorema di Lions sulle equazioni differenziali astratte del secondo ordine. Rend. Sem. Mat. Padova **34**, 224—241 (1964).

Treves, F.

1. Relations de domination entre opérateurs différentiels. Acta Math. **101**, 1—139 (1959).
2. Problèmes de Cauchy et problèmes mixtes en théorie des distributions. J. Anal. Math. Israel **7**, 105—187 (1959).
3. Opérateurs différentiells hypoelliptiques. Ann. Inst. Fourier **9**, 2—73 (1959).
4. Distributions with values in variable spaces. Anais Acad. Bresileira de Ciencias **38** 1, 5—9 (1966).
5. Linear partial differential equations with constant coefficients, New York: Gordon & Breach 1966.

Trotter, H. F.

1. Approximation of semigroups of operators. Pacific J. Math. **8**, 887—919 (1958).

Uraltzeva, N. N., Ladyženskaya, O. A.
Cf. Ladyženskaya-Uraltzeva.

Uraltzeva, N. N., Ladyženskaya, O. A., Solonnikov, V. A.
Cf. Ladyženskaya-Solonnikov-Uraltzeva.

Vaghi, C.

1. Un complemento alla regolarizzazione del problema misto per l'equazione iperbolica. Rend. Ist. Lombardo **100**, 685—705 (1966).

Vallee, A.

1. Un problème de contrôle optimum dans certaines équations différentielles d'évolution. Ann. Sc. Norm. Sup. Pisa **20**, 25—30 (1966).

Vilenkin, N. Ya., Gelfand, I. M.
Cf. Gelfand-Vilenkin.

Vishik, I. M.

1. Sur les systèmes fortement elliptiques d'équations différentielles. Mat. Sbornik **29**, 615—676 (1951).

2. Sur les problèmes aux limites généraux pour les équations aux dérivées partielles elliptiques. Trudy Moscow Mat. Obsc. **1**, 187—246 (1952) [Amer. Math. Soc. Transl. (2) **24**, 107—172 (1963)].
3. Problèmes aux limites pour les équations elliptiques dégénérant à la frontière. Mat. Sbornik **35**, 513—568 (1954) [Amer. Math. Soc. Transl. (2) **35**, 15—78 (1964)].
4. Le problème de Cauchy pour les équations avec coefficients opérateurs. Mat. Sbornik **39** (81), 51—148 (1956) [Amer. Math. Soc. Transl. (2) **24**, 173—278 (1963)].
5. Equations elliptiques de convolution dans un domaine borné et applications, Lecture delivered at the International Congress, Moscow, 1966, Proceedings, 409—420.

Vishik, I. M., Agranovich, M. S.
Cf. Agranovich-Vishik.

Vishik, I. M., Eskin, I. G.

1. Equations de convolution dans un domaine borné. Uspehi Mat. Nauk **20**, No. 3, 90—152 (1965) [Russian Math. Surv. **20**, 85—151 (1965)].
2. Equations de convolution dans un domaine borné dans des espaces avec poids. Mat. Sbornik **69** (111), 65—110 (1966).
3. Equations paraboliques de convolution dans des domaines bornés. Mat. Sbornik **71** (113), 162—190 (1966).
4. Equations elliptiques de convolution dans des domaines bornés et applications. Uspehi Mat. Nauk **22**, 15—76 (1967).
5. Normally solvable problems for elliptic systems of equations in convolution. Mat. Sbornik **74** (116), 326—356 (1967) [Math. U.R.S.S. **3**, No. 3 (1967)].

Vishik, I. M., Ladyženskaya, O. A.

1. Problèmes aux limites pour les équations aux dérivées partielles et certaines classes d'équations opérationnelles. Uspehi Mat. Nauk **11**, No. 6, 41—97 (1956) [Amer. Math. Soc. Transl. (2) **10**, 223—281 (1958)].

Vishik, I. M., Sobolev, S. L.
Cf. Sobolev-Vishik.

Volevich, L. R.

1. Propriétés locales des solutions des systèmes quasi elliptiques. Mat. Sbornik **59** 101, 3—52 (1962).
2. Problèmes aux limites pour systèmes elliptiques généraux. Mat. Sbornik **69** 110, 373—416 (1965).

Volevich, L. F., Panejach, B. P.

1. Certains espaces de fonctions généralisées et théorèmes d'inclusion. Uspehi Mat. Nauk **20**, No. 1, 3—74 (1965) [Russian Math. Surv. **20**, 1—73 (1965)].
2. Values of vector valued distributions and smoothness of semi-group distributions. Bull. Kyushu Inst. Tech. Math. Nat. Sci. **12**, 1—27 (1965).

Walter, W.

1. Über die Euler-Poisson-Darboux Gleichung. Math. Zeit. **67**, 361—376 (1957).

Wang, P. K. C.

1. Control of distributed parameter systems. Advances in Control Systems, ed. by C.T. Leondes, **1**, 75—172 (1964).

Weinstein, A.

1. Etude des spectres des équations aux dérivées partielles. Mémorial Sci. Math., No. 88, Paris: Gauthier-Villars 1937.
2. On the wave equation and the equation of Euler-Poisson. Amer. Math. Soc. Symp. Appl. Math. 1952, vol. 5, 137—148.

Wilcox, C.

1. Mathematical Foundations of Diffraction Theory. Math. Res. Center, Tech. Rep. No. 256, Univ. Wisconsin, 1961.
2. The domain of dependence, inequality and initial-boundary value problems for symmetric hyperbolic systems. Bull. Amer. Math. Soc. **70**, 149—154 (1964).

Yoshinaga, K.

1. Ultra distributions and semi-group distributions. Bull. Kyushu Inst. Tech. Math. Nat. Sci. **10**, 1—24 (1963).

Yosida, K.

1. An operator theoretical integration of the wave equation. J. Math. Soc. Japan **8**, 79—92 (1956).
2. Functional Analysis, Grundlehren math. Wissenschaften, Bd. 123, Berlin/Heidelberg/New York: Springer 1965.
3. Time dependent evolution equations on a locally convex space. Math. Annalen **162**, 83—86 (1965).

Zaidmann, S.

1. Un teorema di esistenza globale per alcune equazioni differenziali astratte. Ricerche di Mat. **13**, 56—69 (1964).
2. Convexity properties for weak solutions of some differential equations in Hilbert spaces. Canad. J. Math. **17**, 802—807 (1965).

Zolezzi, T.

1. Teoremi di esistenza nella teoria dei controlli ottimi in dimensione infinita. Rend. Sem. Mat. Univ. Padova **61**, 72—96 (1968).

Additional Bibliography

Chapter 4

For parabolic equations of higher order and related boundary value problems cf.:

Lagnese, J. E.

1. On equations of evolution and parabolic equations of higher order in t. J. Math. Anal. Appl. **32**, 15—37 (1970).

Lewis, J. E.

1. A Poisson integral formula for solutions of parabolic partial differential equations. J. Math. Anal. Appl., **26**, 479—511 (1969).

Piriou, A.

1. Une classe d'opérateurs pseudo-différentiels du type de Volterra; application aux problèmes aux limites paraboliques pour les opérateurs différentiels. C. R. Acad. Sic. Paris **269** A, 692—695 (1969).
2. Problèmes aux limites généraux pour des opérateurs différentiels paraboliques dans un domaine borné. Ann. Inst. Fourier **21**, 59—78 (1971).

For L^p-theory of parabolic equations cf.:

Fabes, E. B., Riviere, N. M.

1. L^p-estimates near the boundary for solutions of the Dirichlet problem. Ann. Sc. Norm. Sup. Pisa **24**, 491—553 (1970).

Kumano-Go, H., Nagase, M.

1. L^p-theory of pseudo-differential operators. Proc. Japan Acad. **46**, 138—142 (1970).

For applications to parabolic equations of results for abstract differential equations (cf. also Chapter 3) we refer to:

Baiocchi, C.

1. Teoremi di regolarità per le soluzioni di equazioni differenziali astratte, Ist. Naz. Alta Mat., Symposia Math., Roma, 1970.
2. Problemi misti per l'equazione del calore, Rend. Sem. Mat. Fis., Milano, 1970.
3. Soluzione debole dei problemi ai limiti per le equazioni paraboliche del tipo del calore. Rend. Ist. Lombardo Sc. Lett. A **103** 704—726 (1969).

Bardos, C.

1. A regularity theorem for parabolic equations. J. Funct. Anal. **7**, 311—322 (1971).

Carrol, R.

1. Abstract methods in Partial Differential Equations, New York: Harper & Row 1969.

For parabolic equations with discontinuous coefficients cf.:

Baiocchi, C.

1. 2., 3., already quoted in this additional bibliography, and

Campanato, S.

1. Sul problema di Cauchy-Dirichlet per equazioni paraboliche del secondo ordine non variazionali, a coefficienti discontinui. Rend. Sem. Mat. Univ. Padova **41**, 153—163 (1968).

Marino, M.

1. Sull'esistenza delle soluzioni deboli dei problemi al contorno per operatori parabolici. Le Matematiche **13**, 387—407 (1968).

Spagnolo, S.

1. Sul limite delle soluzioni di problemi di Cauchy relativi all'equazione del calore. Ann. Sc. Norm. Sup. Pisa **21**, 657—699 (1967).
2. Sulla convergenza di soluzioni di equazioni paraboliche ed ellittiche. Ann. Sc. Norm. Sup. Pisa **22**, 571—597 (1968).

For parabolic equations with "delay" terms cf.:

Artola, M.

1. Sur les perturbations des équations d'évolution; application à des problèmes de retard. Ann. Sci. Ec. Norm. Sup. **4**, **2**, 137—253 (1969).

Comincioli, V.

1. Problemi periodici relativi a equazioni d'evoluzione paraboliche con termini di ritardo, . . . Rend. Ist. Lombardo Sc. Let. A **104**, 356—381 (1970).
2. Ulteriori osservazioni sulle soluzioni del problema periodico per equazioni paraboliche lineari con termini di perturbazione. Rend. Ist. Lombardo Sc. Let. A **104**, 726—735 (1970).

For boundary value problems in a non-cylindrical open set (cf. problem 17.8) and for Cauchy-mixed problems (cf. problem 17.16) see:

Baiocchi, C.

2. already quoted in this additional bibliography.

For non-negative solutions of linear parabolic equations and their "traces" cf.:

Aronson, D. G.

1. Non negative solutions of linear parabolic equations. Ann. Sc. Norm. Sup. Pisa **22**, 667—694 (1968).
2. Non-negative solutions of linear parabolic equations; an addendum. Ann. Sc. Norm. Sup. Pisa **23** (1971).

Many new results have been obtained for quasi-elliptic equations; see for instance:

Barozzi, C.

1. Un problema al contorno non omogeneo in un dominio angoloso per equazioni fortemente quasi-ellittiche, I and II. Rend. Sem. Mat. Univ. Padova **44** (1970).

Kannai, Y.

1. On the asymptotic behavior of resolvent Kernels, spectral functions and eigen-values of semi-elliptic systems. Ann. Sc. Norm. Sup. Pisa **23**, 563 to 634 (1969).

Pini, B.

1. Sulla rappresentazione delle soluzioni delle equazioni quasi-ellittiche in una regione angolosa, Ann. Mat. Pura Appl. **4** 80, 359—372 (1968).

Troisi, M.

1. Un problema al contorno per una classe di equazioni lineari quasi-ellittiche. Ricerche di Mat. **17**, 263—285 (1968).

Uspenskii, S. V.

1. Estimations on the infinity of solutions in one class of semi-elliptic equations. Rend. Acc. Naz. Lincei **8** 56, 641—645 (1969).

For linearized equations of water waves (cf. problem 17.12) cf.:

Friedman, A., Shinbrot, M.

1. The initial value problem for the linearized equations of water waves, II. J. Math. Mech. **18**, 1177—1193 (1969).

Chapter 5

There has been a large amount of effort concerned with the solution of general boundary value problems for hyperbolic operators (cf. Problem 14.7), cf.:

Agemi, R., Shirota, T.

1. On necessary and sufficient conditions for L^2-well-posedness of mixed problems for hyperbolic equations. J. Fac. Sci. Hokkaido Univ., Ser. I **21**, No. 2, 133—151 (1970).

Ikawa, M.

1. A mixed problem for hyperbolic equations of second order with a first order derivative condition. Pub. Res. Inst. Math. Sci., Kyoto Univ., Ser. A **5**, No. 2, 119—147 (1969).
2. On the mixed problem for hyperbolic equations of second order with the Neumann boundary condition. Osaka J. Math. **7**, No. 1, 203—223 (1970).
3. Mixed problem for the wave equation with an oblique derivative boundary condition, to appear.

Inone, A.

1. On the mixed problem for the wave equation with an oblique boundary condition. J. Fac. Sci., Univ. of Tokyo **16**, 313—329 (1970).

Kreiss, H. O.

1. Initial boundary value problems for hyperbolic systems. Commun. Pure Appl. Math. **23**, 277—298 (1970).

Sadamatsu, T.

1. On mixed problems for hyperbolic systems of first order with constant coefficients. J. Math. Kyoto Univ. **9**, No. 3, 399—361 (1969).

Sakamoto, R.

1. Mixed problems for hyperbolic equations I. J. Math. Kyoto Univ. **10**, No. 2, 349—373 (1970).
2. Mixed problems for hyperbolic equations II. J. Math. Kyoto Univ. **10**, No. 3, 375—401 (1970).

Shirota, T.

1. The wave propagations of mixed problems for hyperbolic equations, to appear.

Shirota, T., Agemi, R.

1. On certain mixed problems for hyperbolic equations of higher order III. Proc. Japan Acad. **45**, No. 10, 854—858.

We refer in particular to the work of R. Sakamoto listed above. Let us also point out the problem of *the oblique derivative* for, say, the wave operators (cf. Problem 14.18), where progress has been made: existence and uniqueness of very general solutions in the sense of Volume 3 have been proved by

Chazarain, R.

1. Problèmes de Cauchy abstraits et applications à quelques problèmes mixtes. J. Funct. Anal., **7**, 386—446 (1971).

For boundary value problems for systems of first order equations cf.:

Agranovich, M. S.

1. Boundary value problems for systems of first order pseudo-differential operators. Uspehi Mat. Nauk **24**, 59—128 (1969) [Russian Math. Surv. **24**, 61—126 (1969)]

Friedrichs, K. O., Lax, P. D.

1. On symmetrizable differential operators. Proc. Symp. pure math., Amer. Math. Soc. **10**, 128—137 (1968).

The Dirichlet problem for positive hyperbolic operators has been considered by:

Authier, M.

1. Problème de Dirichlet pour des opérateurs hyperboliques de type positif. Ann. Sc. Norm. Sup. Pisa (to appear).

Coerciveness for non elliptic systems and applications for instance to Maxwell's equations are considered in:

Wilcox, C. H.

1. Coerciveness inequalities for non elliptic systems of partial differential equations. Ann. Mat. Pura Appl. (to appear).

For L^p-theory for wave and Schroedinger's equations (Cauchy problem) cf.:

Lanconelli, E.

1. Valutazioni in L_p della soluzione del problema di Cauchy per l'equazione di Schroedinger. Boll. U.M.I. **4** 1, 591—609 (1968).
2. Valutazioni in B_p^0 della soluzione del problema di Cauchy per l'equazione delle onde. Boll. U.M.I. **4** 1, 780—790 (1968).

For scattering theory cf.:

Kato, T.

1. Scattering theory and perturbation of continuous spectra. Proc. Int. Congress of Math., Nice, 1970.

Phillips, R.

1. Scattering theory for the wave equation in the exterior of an obstacle. Proc. Int. Congress Math., Nice, 1970.

For the theory of lacunas for hyperbolic operators cf.:

Atiyah, M. F., Bott, R., Gärding, L.

1. Lacunas for hyperbolic differential operators with constant coefficients, I. Acta Math. **124**, 109—189 (1970).

Gärding, L.

1. Lacunas for hyperbolic differential operators. Proc. Intern. Congress of Math., Nice, 1970.

Finally we must quote the recent generalization of the theory of pseudo-differential operators developed by Hörmander (cf. also Maslov, Eskin, Egorov, Nirenberg-Treves) which also contains hyperbolic operators and which seems to be also useful for boundary value problems for hyperbolic operators:

Egorov, Yu. V.

1. On canonical transformations of pseudo-differential operators. Uspehi Mat. Nauk **25**, 235—236 (1969).
2. On non degenerate hypoelliptic pseudo-differential opérators. Dokl. Akad. Nauk **186**, 1006—1007 (1969).

Eskin, G. I.

1. The Cauchy problem for hyperbolic systems in convolution. Mat. Sbornik **74** (116), 262—297 (1967).
2. Pseudo differential systems of equations with simple real characteristics. Mat. Sbornik **77**, 174—200 (1968).

Hörmander, L.

1. On the singularities of solutions of partial differential equations. Conf. on Funct. Anal. and related Topics, Tokyo 1969, 31—40.
2. The Calculus of Fourier Integral Operators. Acta Math. **127**, 79—183 (1971).

Maslov, V. P.

1. Perturbation theory and asymptotic methods, Moscow 1965.

Nirenberg, L., Treves, F.

1. On local solvability of linear partial differential equations. Part I: Necessary conditions, Part II: Sufficient conditions. Comm. pure appl. math. **23**, 1–38 and 459–510 (1970).

Treves, F.

1. Hamiltonian fields, bicharacteristic strips in relation with existence and regularity of partial differential equations, Proc. Intern. Congress of Math., Nice, 1970.

Chapter 6.

For the optimal control of systems governed by *stochastic* partial differential equations, cf.:

Bensoussan, A.

Filtrage optimal des systèmes linéaires, Paris: Dunod 1971.
and the bibliography of this book.

For systems with delays, let us mention:

Delfour, M. C., Mitter, S. K.

Hereditary differential systems with constant delays, I: General Case, Univ. de Montreal, 1970, and the bibliography of this work.

For systems with several "controllers" and relations to game theory cf.

Lemaire, B.

Problèmes min-max et applications au controle optimal de systèmes gouvernés par des équations aux derivées partielles, Thesis, Paris 1970.

cf. also the work of Bensoussan, Yvon and Lions, to appear, a short account of it being given in a survey paper by Lions for the IFAC Symposium, Banff, June 1971.

For optimal control theory in Banach spaces cf.

Friedman, A.

1. Optimal control in Banach spaces. J. Math. Anal. and Appl. **19**, 35–55 (1967).
2. Optimal control in Banach spaces with fixed end-points. J. Math. Anal. and Appl. **24**, 161–181 (1968).

And for existence theorems cf.:

Fleming, W. H.

1. Optimal continuous-parameter stochastic control. S.I.A.M. Review **11**, 470–509 (1969).

Zolezzi, T.

1. Teoremi di esistenza per problemi di controllo ottimo retti da equazioni ellittiche o paraboliche. Rend. Sem. Mat. Univ. Padova **44**, 155–175 (1970).

721/3/72 – III/18/203

Die Grundlehren der mathematischen Wissenschaften in Einzeldarstellungen mit besonderer Berücksichtigung der Anwendungsgebiete

Eine Auswahl

---

147. Lamotke: Semisimpliziale algebraische Topologie. DM 56,—
148. Chandrasekharan: Introduction to Analytic Number Theory. DM 28,—
149. Sario/Oikawa: Capacity Functions. DM 96,—
150. Iosifescu/Theodorescu: Random Processes and Learning. DM 68,—
151. Mandl: Analytical Treatment of One-dimensional Markov Processes. DM 36,—
152. Hewitt/Ross: Abstract Harmonic Analysis. Vol. 2: Structure and Analysis for Compact Groups. Analysis on Locally Compact Abelian Groups. DM 140,—
153. Federer: Geometric Measure Theory. DM 118,—
154. Singer: Bases in Banach Spaces I. DM 112,—
155. Müller: Foundations of the Mathematical Theory of Electromagnetic Waves. DM 58,—
156. van der Waerden: Mathematical Statistics. DM 68,—
157. Prohorov/Rozanov: Probability Theory. DM 68,—
159. Köthe: Topological Vector Spaces I. DM 78,—
160. Agrest/Maksimov: Theory of Incomplete Cylindrical Funktions and Their Applications. DM 84,60
161. Bhatia/Szegö: Stability Theory of Dynamical Systems. DM 58,—
162. Nevanlinna: Analytic Functions. DM 76,—
163. Stoer/Witzgall: Convexity and Optimization in Finite Dimensions I. DM 54,—
164. Sario/Nakai: Classification Theory of Riemann Surfaces. DM 98,—
165. Mitrinović/Vasić: Analytic Inequalities. DM 88,—
166. Grothendieck/Dieudonné: Eléments de Géométrie Algébrique I. DM 84,—
167. Chandrasekharan: Arithmetical Functions. DM 58,—
168. Palamodov: Linear Differential Operators with Constant Coefficients. DM 98,—
170. Lions: Optimal Control Systems Governed by Partial Differential Equations. DM 78,—
171. Singer: Best Approximation in Normed Linear Spaces by Elements of Linear Subspaces. DM 60,—
172. Bühlmann: Mathematical Methods in Risk Theory. DM 52,—
173. F. Maeda/S. Maeda: Theory of Symmetric Lattices. DM 48,—
174. Stiefel/Scheifele: Linear and Regular Celestial Mechanics. Perturbed Two-body Motion—Numerical Methods—Canonical Theory. DM 68,—
175. Larsen: An Introduction to the Theory of Multipliers. DM 84,—
176. Grauert/Remmert: Analytische Stellenalgebren. DM 64,—
177. Flügge: Practical Quantum Mechanics I. DM 70,—
178. Flügge: Practical Quantum Mechanics II. DM 60,—
179. Giraud: Cohomologie non abélienne. DM 109,—
180. Landkoff: Foundations of Modern Potential Theory. DM 88,—
181. Lions/Magenes: Non-Homogeneous Boundary Value Problems and Applications I. DM 78,—
182. Lions/Magenes: Non-Homogeneous Boundary Value Problems and Applications II. DM 58,—
184. Rosenblatt: Markov Processes. Structure and Asymptotic Behavior. DM 68,—
185. Rubinowicz: Sommerfeldsche Polynommethode. DM 56,—
186. Wilkinson/Reinsch: Handbook for Automatic Computation II, Linear Algebra. DM 72,—
187. Siegel: Lectures on Celestial Mechanics. DM 78,—
188. Warner: Harmonic Analysis on Semi-Simple Lie Groups I. In preparation
189. Warner: Harmonic Analysis on Semi-Simple Lie Groups II. In preparation

MIX
Papier aus verantwortungsvollen Quellen
Paper from responsible sources
FSC® C105338

If you have any concerns about our products,
you can contact us on
ProductSafety@springernature.com

In case Publisher is established outside the EU,
the EU authorized representative is:
Springer Nature Customer Service Center GmbH
Europaplatz 3, 69115 Heidelberg, Germany

Printed by Libri Plureos GmbH
in Hamburg, Germany